IMAGES
of America

NEW MEXICO
SPACE TRAIL

ON THE COVER: Apollo astronauts David Scott (left) and James Irwin (right) ride in the Geologic Rover (Grover) on a geologic training trip to Taos, New Mexico, September 9–10, 1971. Since the original Lunar Rover was designed for the moon's low gravity, it would collapse under the weight of fully suited astronauts. Engineers at the US Geologic Survey created a near-copy that could be used by astronauts on Earth during training. (Courtesy of NASA.)

IMAGES
of America

NEW MEXICO
SPACE TRAIL

Joseph T. Page II

ARCADIA
PUBLISHING

Published by Arcadia Publishing
Charleston, South Carolina

Printed in the United States of America

Library of Congress Control Number: 2013935783

For all general information, please contact Arcadia Publishing:
Telephone 843-853-2070
Fax 843-853-0044
E-mail sales@arcadiapublishing.com
For customer service and orders:
Toll-Free 1-888-313-2665

Visit us on the Internet at www.arcadiapublishing.com

To Lt. Col. Wayne Otto Mattson (1929–2012)—you have taken the story threads of New Mexico's space history and helped weave a beautiful tapestry of heritage, pride, and respect. (Photograph by Ron Keller, courtesy of New Mexico Museum of Space History.)

CONTENTS

FOREWORD

Even as a relative newcomer to the state of New Mexico, it did not take me long to discover the joy of hiking here. The state takes great pride in the fascinating and extensive trails through its stunning canyons, mountains, desert plains, and mesas. These walking and biking trails are constantly being developed and improved by dedicated volunteers, as well as federal and state employees. It's not surprising, then, that a number of years ago, a couple employees of the New Mexico Museum of Space History started thinking about linking together the plethora of space history sites in the state and the idea of a "trail" seemed the logical choice to lay out their vision.

Some quick research and discussion led to the initial inclusion of 24 sites across the state. As you would do when blazing a new trail, the points were laid out and dutifully numbered so that the visitor could follow their way through the amazing history of space exploration in New Mexico. Prehistoric star-gazing sites led to locations where the first rockets were launched, which led to sites where the limits of human endurance were tested, which led to sites where the first men to walk on the moon trained, which led to sites where today's entrepreneurs are preparing to send everyday travelers on journeys to the stars. Already it seemed that there was an amazing collection of historical locales for our guests to learn about, visit, and enjoy . . . but that was just the beginning.

Men and women from across the spectrum of space exploration and scientific research have been enchanted by New Mexico and their stories play out as you read about them or visit many of these historic sites. Dr. Robert Goddard, Werner von Braun, Colonel John Paul Stapp, and Neil Armstrong are just a few of the individuals who left their mark on the state and blazed a trail to the stars. Exploring the Space Trail can take you back in history to 100 BCE and the rock alignments at Wizard's Roost in the Sacramento Mountains that were used as a calendar, similar to the much larger Stonehenge in England. You can learn about the ranchers displaced by the U.S. Army in 1945 to create the 2,671,000 acre White Sands Proving Ground; a site that later became known as the White Sands Missile Range and is nicknamed the "Birthplace of Space." You can walk in the footsteps of the Apollo astronauts when you visit sites like the Zuni Salt Lake and the Rio Grande Gorge area where they learned to walk—and drive—on the moon. And, as you follow this trail, you will also look to the future at sites like Spaceport America where today's entrepreneurs are getting ready to use the world's first dedicated spaceport to launch spacecraft that will perform experimental research for NASA and take individual travelers on their own personal journeys to the stars.

In the 12 years since its inception, the trail has gone from the original 24 sites to the current total of 52—and the list should expand by at least another half dozen over the next two years. As you leaf through the pages of this book, you will get a taste of the history of these sites and revel in the imagery of the explorers who have gone before you. Relax, enjoy, learn, contemplate the grandeur and excitement of sites you may have just visited . . . and see what new adventures you can enjoy as you blaze your own personal journey on New Mexico's Space Trail.

—Chris Orwoll, director of the New Mexico Museum of Space History

ACKNOWLEDGMENTS

This project could not have been completed without the backing of the New Mexico Museum of Space History (NMMSH). The team of dedicated professionals includes director Chris Orwoll, curator George House, educational specialist Michael Shinabery, marketing director Cathy Harper, Jim Mayberry, and photographer Ron Keller. It was through collaboration with the NMMSH that I met the indomitable Lt. Col. Wayne O. Mattson.

I'd also like to thank the following for their assistance: at New Mexico State University (NMSU), Pres. Barbara Couture, professors Jon Holtzman, Nancy Chanover, and Beth O'Leary, and chief photographer Darren Phillips; at New Mexico Tech, Alisa Shtromberg and Edie Steinhoff; at the University of New Mexico, Prof. Rich Rand and Jon Lewis; at the New Mexico School for the Blind and Visually Impaired, superintendent Linda Lyle and Lee Rohovec; at Sandia National Laboratories, Randy Montoya; at Los Alamos National Laboratory, Linda Deck; at Tzec Maun Foundation, Ron Wodaski; at Valles Caldera National Preserve, Dr. Anastasia Steffen; at the 579th Strategic Missile Squadron Association, Fred Mortimer; at the Walker Aviation Museum Foundation, Judy Armstrong; at the Roswell Museum and Art Center, Laurie Rufe and Caroline Brooks; at the Columbia Scientific Balloon Facility, Danny Ball and Darla Cook; at the National Museum of Nuclear Science and History, Jennifer Hayden; at the National Radio Astronomy Observatory, Bill Saxton; at Spaceport America, David Wilson; New Mexico state representative Dennis Kintigh; at White Sands Test Facility, Cheerie Patneaude and Robert Cort; at the Philmont Museums, David E. Werhane; and at the White Sands Missile Range Museum, Darren Court.

I am indebted to Prof. Pete Eidenbach (NMSU-Alamogordo)—first for discovering Wizard's Roost/ Wally's Dome and second for the pictures. Thanks to Stacia Bannerman for her editorial guidance and patience. Also, love to my wife and children for being my own personal cheerleading squad.

Thanks to my baristas for caffeinated fuel: Nadeja, Mykayla, Kirsten, Isolde "Isa," Stacey, Kelly, Grace, Buddy, Lauren, and Airon.

Finally, infinite thanks to a pair of resolute alpha females. Without their assistance, this book would not exist. First, thanks to Melissa Wilde and the caffeine-fueled stories of the Tularosa Basin. And thanks to Capt. Erin E. Gaberlavage—she has the heart of a warrior and the soul of an artist.

All uncredited photographs belong to the author.

INTRODUCTION

The Mescalero Apache people of southeastern New Mexico have a very strong connection with the universe through ethnoastronomy. Described by Morris Opler, the "dean of Apachean scholarship," the Mescalero "live the sky" on a daily basis. So strong are their beliefs that, as with other Native American tribes in the Southwest, characteristics of their culture have remained through present-day New Mexico symbols. The individual threads of the New Mexico Space Trail have similarly permeated through almost every inch of this southwestern state.

The roots of the trail start with mankind's emergence in the Southwest over 11,000 years ago. The Clovis culture, named for their discovery near Clovis, New Mexico, were a group of hunter-gatherers. While primarily hunters, they did show evidence of rudimentary agricultural knowledge as gatherers. This next step in evolutionary anthropology, advanced knowledge of agriculture, would become a founding reason for the earliest sites on the Space Trail.

Chaco Canyon was a great center of culture for the Ancient Pueblo peoples. Collection of foodstuffs and storage made it imperative that the peoples knew when spring planting and autumn harvest would occur. The key to this schedule became the solstices and equinoxes. The culture's complex knowledge of celestial events is documented in evidence found at Chaco Canyon, seemingly from generations of astronomical observations.

Other archaeoastronomy sites around the state provide incontrovertible evidence of ancient humans' knowledge of the universe with practical applications. Two sites in the Sacramento Mountains dubbed Wally's Dome and Wizard's Roost form two nodes in an intricate network of solstice observation posts. The draw to New Mexicans of viewing the heavens has continued into the present day with technologically advanced instruments. Astronomical observatories are present at all major educational institutions, such as New Mexico State University (NMSU), New Mexico Tech (NMT), and the University of New Mexico (UNM). The Karl G. Jansky Very Large Array (VLA), run by the National Radio Astronomy Observatory (NRAO), is splayed out on the Plains of San Augustin, in the west central part of the state. Observatories atop the Sacramento Mountains monitor the solar system's favorite thermonuclear heat source, the sun.

Breaking the stereotypical paradigm of astronomy research, two entities located within the Sacramento Mountains have taken an entrepreneurial spin on astronomy education for the masses. The Tzec Maun Foundation and Earthrise Institute have created an impressive collection of telescopes and very desirable locations for gazing at the heavens. Tzec Maun's telescopes are accessible through the Internet, allowing anyone with access to steer the scopes.

Once humankind's technical ability to reach the stars synchronized with the universe's entrancing draw, the first steps to space began—right here in New Mexico. Highlights along the trail, such as Dr. Robert H. Goddard's rocket experiments, V-2 missile launches at White Sands Missile Range (WSMR), medical tests for Mercury astronauts, training of the first chimpanzee in space, and the creation of the world's first commercial spaceport, are best summed up by New Mexico's state motto, *Crescit Eundo*—"It grows as it goes." This succinct motto is a perfectly apt description of the continuing legacy of New Mexico's contribution to space exploration.

One

ORIGINS OF THE SPACE TRAIL

As travelers cross the New Mexico state line, they are welcomed by the words "Welcome to the Land of Enchantment" on highway signs along the edge of the interstate. Additionally, visitors are also welcomed into New Mexico's storied history by various historic highway markers commemorating famous dates within the state's history or prominent geographic features within view of the marker. The markers give pause to the new visitors—instead of speeding through the Southwest at breakneck speeds, the markers encourage one to reflect on the activities of the past.

The origins of the New Mexico Space Trail took seed in the New Mexico Department of Cultural Affairs Historic Preservation Division's simple, yet effective, tourism program: the highway historic markers. Starting in 1935, New Mexico tourism gurus used simple signs mounted inside wood frames to entice highway travelers to stop and admire the sites while spending some money on souvenirs and food. These original markers were simple, hand-cut wood with the informative text scrawled onto them.

In the intervening decades, the Highway Historic Trail grew to over 300 signs around the state, and it is viewed by millions of highway travelers per year. On one of these highways, New Mexico Museum of Space History curator George House passed by a historical marker and asked, "There are no space-related historical markers anywhere in the state. Why not combine the two and have a space trail?"

Imagining roadside markers denoting the state's space legacy and promoting tourism, House envisioned a family-friendly summer vacation driving through the state and visiting each site. Budget constraints, however, precluded the state erecting new roadside markers. Through due diligence, the Space Trail idea didn't die, it flourished. Using information he researched for a 1990 presentation to a historical society, House developed maps outlining the trail, while the museum's savvy research and marketing team expanded on the educational efforts of the trail's legacy.

In 2010, working with the museum, New Mexico representative Dennis Kintigh (R-Roswell) introduced a memorial bill designating the trail, completing a task that had been compiled over the span of two decades.

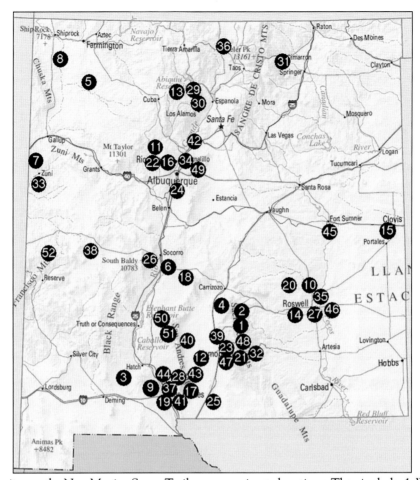

All the sites on the New Mexico Space Trail are approximate locations. They include: 1. Wizard's Roost; 2. Wally's Dome; 3. Tenuco Observatory; 4. Three Rivers Petroglyphs; 5. Chaco Canyon; 6. Scholle Crab Nebula Supernova Petroglyph; 7. Zuni Pueblo Crab Nebula Supernova Petroglyph; 8. Sand Paintings, Navajo Reservation; 9. New Mexico State University; 10. Robert Goddard Rocket Research Site; 11. Kirtland Air Force Base; 12. Holloman Air Force Base; 13. Los Alamos National Laboratory; 14. Walker Air Force Base; 15. Cannon Air Force Base; 16. Sandia National Laboratory; 17. White Sands Missile Range; 18. Trinity Site; 19. NMSU Physical Science Laboratory; 20. Alleged UFO Crash Sites; 21. Sacramento Peak, Sunspot, New Mexico; 22. University of New Mexico Campus Observatory; 23. New Mexico School for the Blind and Visually Impaired; 24. Lovelace Medical Center; 25. McGregor Range; 26. New Mexico Tech; 27. Atlas Missile Silos; 28. White Sands Test Facility; 29. Bradbury Science Museum; 30. Valles Caldera National Preserve; 31. Philmont Boy Scout Ranch; 32. Tzec Maun Observatory; 33. Zuni Salt Lake; 34. National Museum of Nuclear Science and History; 35. Roswell Museum and Art Center; 36. Taos/Rio Grande Gorge; 37. Clyde W. Tombaugh Campus Observatory; 38. Very Large Array Radio Telescope; 39. New Mexico Museum of Space History; 40. White Sands Space Harbor; 41. Tortugas Mountain Planetary Observatory; 42. University of New Mexico's Capilla Peak Observatory; 43. High Energy Laser Systems Test Facility; 44. Tracking and Data Relay Satellite System; 45. Columbia Scientific Balloon Facility; 46.International UFO Museum and Research Center; 47. Apache Point Observatory, Sunspot, New Mexico; 48. Comet Hale Bopp Codiscovered, Cloudcroft, New Mexico; 49. New Mexico Museum of Natural History & Science/Planetarium; 50. Spaceport America; 51. Virgin Galactic International Headquarters; and 52. Magdalena Ridge Observatory.

The stylized logo of the New Mexico Space Trail includes the Zia sun symbol, prominently displayed on the New Mexico state flag. Four is the sacred number of the Zia people, and the symbol is composed of a circle from which four points radiate. The lines represent the four cardinal directions; seasons of the year; the day, with the sunrise, noon, evening, and night; and life with its four divisions—childhood, youth, manhood, and old age. (Courtesy of NMMSH.)

This is an example of the official New Mexico highway scenic historic markers posted around the state. This particular sign details the disappearance of Albert J. Fountain and his son Henry within the confines of what is now known as White Sands National Monument. George House, New Mexico Museum of Space History curator, passed this marker on the way to a conference. His curiosity came to fruition as the New Mexico Space Trail.

OFFICIAL SCENIC HISTORIC MARKER

Disappearance of Albert J. Fountain and his son Henry

Albert Jennings Fountain was a Civil War veteran, New Mexico legislator and prominent lawyer. On Febuary 1, 1896 Fountain and his eight–year–old son, Henry, were traveling home to Mesilla from Lincoln. They carried grand jury indictments against cattle rustlers. Both disappeared at Chalk Hill, and their bodies were never found. In 1899 Oliver Lee and James Gililland were tried for their murder. Both were acquitted.

The reverse of the Albert J. Fountain disappearance sign includes a detailed map, included on all New Mexico historic highway markers. The map gives tourists a brief peek at other historic attractions in the nearby area. Three Space Trail sites are prominently listed here: Three Rivers Petroglyphs, Sacramento Peak Solar Observatory, and the NMMSH.

For the first two decades of its existence, the museum was known as the International Space Hall of Fame. Changing names twice in the next two decades, the NMMSH still retains the International Space Hall of Fame as a global roll call of individuals who have dedicated their lives to the furthering of space exploration. (Photograph by Ron Keller, courtesy of NMMSH.)

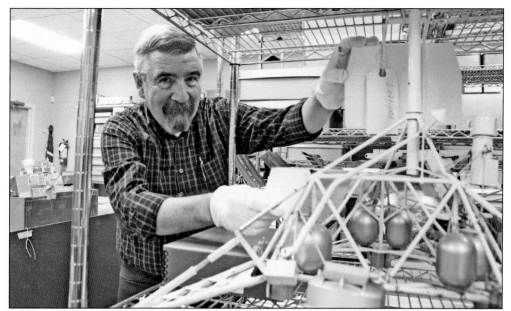

Museum curator George House handles space artifacts at the NMMSH's Hubbard Space Science building. A native New Mexican, House's personal journey took him from the small town of Silver City into service with the US Navy as a corpsman. Returning to the state and earning undergraduate and graduate degrees in history, he became head curator at the NMMSH in 1985. (Photograph by Ron Keller, courtesy of NMMSH.)

The New Mexico Museum of Space History team celebrates the New Mexico Space Trail. From left to right are Noel Romero, information technology systems manager; Michael Shinabery, educational specialist; Chris Orwoll, NMMSH director; George House, museum curator; Cathy Harper, marketing director; and Jim Mayberry, assistant museum curator. Keeping the trail's information up to date for tourists, educators, and students alike is a full-time job but looks easy with this team of dedicated educators and specialists. (Photograph by Ron Keller, courtesy of NMMSH.)

The giant glinting cube-shaped building of the New Mexico Museum of Space History is viewable from over 50 miles away across the Tularosa Basin. The building cost $1.8 million to complete and officially opened on October 5, 1976. Almost 80 feet high and 60 feet on a side, the building was originally envisioned more as an educational facility than tourist attraction. (Photograph by Ron Keller, courtesy of NMMSH.)

An east-facing view of the NMMSH campus shows the John P. Stapp Air & Space Park. Displays include tracking telescopes and telemetry antennas from White Sands Missile Range, an F-1 rocket engine from the Saturn moon rocket, Colonel Stapp's Sonic Wind No.1 rocket sled, a Lance battlefield missile, a Minitrack satellite antenna, an Aerobee sounding rocket, V-2 missile debris, and a Nike Ajax air-defense missile launcher. (Photograph by Ron Keller, courtesy of NMMSH.)

A west-facing view of the John P. Stapp Air & Space Park overlooks the Tularosa Basin. The park is free and open to visitors seven days a week. One prize display within the park is a Little Joe II space launch vehicle, seen off to the right. Sometimes mistaken for a Saturn I rocket, the rocket tested the Apollo launch escape system and command module parachutes at White Sands Missile Range. (Photograph by Ron Keller, courtesy of NMMSH.)

An ingress/egress training simulator of the US Air Force's F-117A Nighthawk stealth fighter resides in the museum's collection. Donated by the 49th Fighter Wing, host unit at the nearby Holloman Air Force Base, the simulator was used to teach Nighthawk pilots how to get into the cockpit (ingress), ground emergency exit (egress), and bailout procedures during flight. Other Air Force artifacts nearby include an X-37B Orbital Test Vehicle mock-up. (Photograph by Ron Keller, courtesy of NMMSH.)

The fourth floor of the museum holds displays ranging from a model of Sputnik to lunar sample containers brought back from the moon. One interesting display, seen near the center of the photograph, showcases two Project Moonwatch handheld telescopes. Envisioned by Harvard astronomer Fred L. Whipple, the project allowed amateur astronomers to track satellites and provide orbital data to government officials during the early days of the Cold War. (Photograph by Ron Keller, courtesy of NMMSH.)

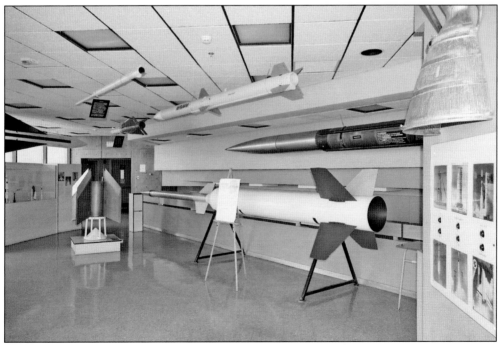

The museum's third floor displays many of the rocket and missile designs used by scientists and engineers. Off to the left is a soundboard allowing users to guess what sounds a particular rocket design makes. (Photograph by Ron Keller, courtesy of NMMSH.)

The Daisy Track, named after the Daisy air rifle, sits outside on NMMSH grounds. The track was air-powered and used to test safety devices at Holloman Air Force Base. The short track length contrasted with Holloman's other track, the 10-mile-long High Speed Test Track. Both, however, made vital and distinctive accomplishments in space research during the early days of the Space Race. (Photograph by Ron Keller, courtesy of NMMSH.)

Beneath the shadow of the International Space Hall of Fame sign lies astrochimp Ham's gravesite. Known as No. 65 before the flight, he was named after the Holloman Aeromedical Research Laboratory, which trained him. Ham died on January 19, 1983, after living out his post-flight years at the National Zoo in Washington, DC, and the North Carolina Zoo in Asheboro. His remains (minus his skeleton) were interred at the NMMSH. (Photograph by Ron Keller, courtesy of NMMSH.)

Astronaut Memorial Garden displays the names of the brave men and women who have sacrificed their lives in the pursuance of spaceflight. The names of the three Apollo 1 astronauts (1967), the crew of the space shuttle *Challenger* (1986), and the names of the *Columbia* crew (2003) are highlighted on the marble obelisk. (Photograph by Ron Keller, courtesy of NMMSH.)

In 2006, Dr. Beth O'Leary, Dr. William Doleman, and New Mexico state historic preservation officer Katherine Slick added Apollo 11's landing site to the state's Archaeological Records Management Section. Entry US LA 2,000,000 forever links Tranquility Base to the NMMSH. This is one of the first efforts to preserve the culture and heritage on the moon and recognizes the first Space Heritage site in the world. (Photograph by Ron Keller, courtesy of NMMSH.)

The only IMAX theater in the state until 2010, the Clyde W. Tombaugh IMAX Theater includes a planetarium with a star projector. The building is a multipurpose venue used by the museum for lectures and planetarium shows as well as for the screening of specially formatted IMAX movies. (Photograph by Ron Keller, courtesy of NMMSH.)

In 2010, Rep. Dennis J. Kintigh introduced New Mexico House Memorial 041 to the state legislature. This bill recognized the contributions to space exploration that occurred within New Mexico. On February 6, 2010, HM041 was passed, with 57 state representatives voting "yes" and eight representatives voting "no." It is unknown what was behind the eight "no" votes. Regardless, the birth certificate for the New Mexico Space Trail was finalized. (Courtesy of the Roswell Rotary Club.)

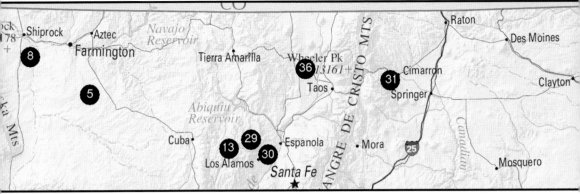

Northern sites on the Space Trail include: 5. Chaco Canyon Sky Watcher's Station; 8. Sand Paintings, Navajo Reservation; 13. Los Alamos National Laboratory; 29. Bradbury Science Museum; 30. Valles Caldera National Preserve; 31. Philmont Boy Scout Ranch; and 36. Taos/ Rio Grande Gorge.

Two

NORTHERN SITES

The mountains and canyons of northern New Mexico have been home to tourists, artisans, and survivalists for centuries. A hard living is earned on the high desert plains, in chilling temperatures and sparseness. The natives who constructed the Chaco Canyon settlements knew the importance of knowing weather and climatic conditions to maximize their harvest. The dwellings and roads throughout the Chaco Canyon complex have been determined to be oriented by the characteristics of celestial influence, including the solstices, and ethnocultural beliefs in the power of the moon and stars.

NASA was quick to recognize that the varied geologic and geographic features of northern New Mexico would greatly benefit the burgeoning astronaut corps during the 1960s. While Pres. John F. Kennedy's goal of landing a man on the moon by the end of the decade was foremost on the astronauts' minds, members of the scientific establishment demanded a return on the billions of dollars spent.

Field exercises held in the Southwest (New Mexico, Arizona, and Colorado) allowed the astronauts, primarily military pilots trained as engineers, to act as geological field observers for Earth-bound scientists. Early investments by this program in places such as the Rio Grande Gorge and Valles Caldera paid many dividends during the "pure" scientific expeditions of Apollo 15, 16, and 17.

The assistance of the Boy Scouts of America's Philmont Scout Ranch was invaluable to the NASA astronauts, as they also spent time performing geologic field activities in the Sangre de Cristo Mountains. The strong connection between the Boy Scouts and NASA was exemplified by the presentation of red wool jac-shirts with a black bull to the astronauts, symbolizing that the wearer participated in a trek at Philmont.

Additionally, many important scientific institutions call northern New Mexico home. Drawing back to his youth, Dr. J. Robert Oppenheimer recommended the Los Alamos Mesa to Gen. Leslie Groves for the Manhattan Project's location. What would later be known as Los Alamos National Laboratory, the location has had a driving hand in the design, security, and stewardship of the nation's nuclear stockpile.

Just outside the great house dwelling at Peñasco Blanco, these pictographs (a handprint, crescent moon, and star) reside at the base of the mesa. Archaeoastronomy researchers believe that these images represent the supernova of 1054, which was visible during daylight hours for 23 days. Present-day astronomers have calculated the moon's phase and supernova's position to match these images precisely. The handprint is believed to be the artist's signature. (Courtesy of James Gordon.)

The Sun Daggers is a rock formation located on Fajada Butte at Chaco Canyon in northern New Mexico. The daggers appear to highlight spiral designs during solstices. Two sun daggers appear at either edge of the spiral during winter solstice, while during the summer solstice, a dagger strikes right through the middle of the design. Erosion and settling of the earth have caused the rocks to move, misaligning the daggers. (Courtesy of NMMSH.)

An overhead view shows Pueblo Bonito ("Beautiful Town") at Chaco Canyon. The complex is shaped like a *D*, with the long straight wall oriented almost perfectly east-to-west. The design features passive solar energy techniques like reflecting light during wintertime and shading from the heat during summer. (Courtesy of NMMSH.)

The kivas at Casa Rinconada have small windows that allow sunlight in during the summer solstice. The alignment of the ruins is nearly precise to the cardinal directions, with the north-south entrances aligned within a third of a degree to true north. (Courtesy of NMMSH.)

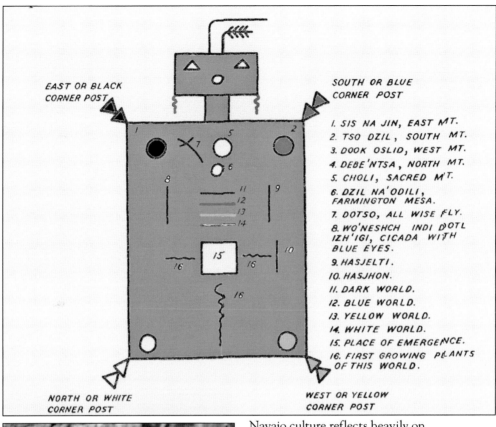

EAST OR BLACK
CORNER POST

SOUTH OR BLUE
CORNER POST

1. SIS NA JIN, EAST MT.
2. TSO DZIL, SOUTH MT.
3. DOOK OSLID, WEST MT.
4. DEBE'NTSA, NORTH MT.
5. CHOLI, SACRED M'T.
6. DZIL NA'ODILI, FARMINGTON MESA.
7. DOTSO, ALL WISE FLY.
8. WO'NESHCH INDI D'OTL IZH'IGI, CICADA WITH BLUE EYES.
9. HASJELTI.
10. HASJHON.
11. DARK WORLD.
12. BLUE WORLD.
13. YELLOW WORLD.
14. WHITE WORLD.
15. PLACE OF EMERGENCE.
16. FIRST GROWING PLANTS OF THIS WORLD.

NORTH OR WHITE
CORNER POST

WEST OR YELLOW
CORNER POST

Navajo culture reflects heavily on ethnoastronomy in its tales of the constellations and cosmology. Navajo sand paintings reflect these stories with ornate designs. Authentic sand paintings are rarely photographed, as they are considered sacred. To create an authentic sand painting solely for viewing would be considered a profane act. (Courtesy of NMMSH.)

Dave Scott (right) is present during field exercises at Taos, New Mexico, on March 11, 1971. The presence of California Institute of Technology's famed geologist Lee Silver (far left, wearing a white hat) illustrated the seriousness NASA attached to field geology training for the moon-bound Apollo astronauts. Apollo 15 was the first "J-mission," with longer three-day stays using the Extended Lunar Module, three scheduled moonwalks, and a Lunar Rover. (Courtesy of NASA.)

Apollo 15 astronauts Jim Irwin (left) and Dave Scott (right) stand on the Rio Grande Gorge's west rim. Irwin holds a traverse map and geologic scoop, while Scott provides verbal descriptions over his radio headset. Apollo 15's crew made one of the most exciting discoveries, finding the "Genesis Rock." Lunar scientists believe the rock was formed in the early stages of the solar system, at least four billion years ago. (Courtesy of NASA.)

A wide-angle photograph shows astronauts Dave Scott (left) and Jim Irwin (right) overlooking the Rio Grande Gorge. Apollo 15's destination, the Hadley-Apennine Mountains, was considered by lunar geologists to be the most geologically diverse region in the entire program. The scientific importance was not lost on NASA planners, who scheduled 18 separate Earth-bound field exercise trips for the Apollo 15 crew before its lunar exploration. (Courtesy of NASA.)

Astronauts wear replicas of their Portable Life Support System (PLSS) backpacks while on an expedition near the Rio Grande Gorge. The mock-ups allowed simulation of any awkward movement or positioning of their sampling equipment they would be faced with on the lunar surface. The PLSS was first tested in orbit by Apollo 9's Russell Schweickart. On Earth, the PLSS weighed 84 pounds; on the lunar surface, it weighed only 14 pounds. (Courtesy of NASA.)

A panoramic view shows Apollo 15 astronauts James Irwin (left) and Dave Scott (right) traversing near the edge of the Rio Grande Gorge in the Grover. The design of the Lunar Rover allowed either passenger to drive, but the mission commander usually took the initiative. A T-handle between the two occupants propelled the vehicle's four independently driven electric wheels. (Courtesy of NASA.)

Dave Scott, commander of the Apollo 15 mission, peers over the Rio Grande Gorge in Taos, New Mexico. As the Apollo program was curtailed, later missions were bolstered to accommodate the scientific community's requests. Apollo 15 was the first mission planned for longer three-day stays and three moonwalks. It was not until Apollo 17, the last manned mission to the moon, that a trained scientist, New Mexico native Harrison "Jack" Schmitt, was sent to the lunar surface. (Courtesy of NASA.)

One factor planners included in the Lunar Rover missions was called the "walkback limit." If the Lunar Rover failed for any reason, the astronauts would have to walk back to the Lunar Module. Moonwalks were designed to start at the far end of the walkback limit and gradually get closer to the lunar lander. Increased Lunar Rover reliability allowed Apollo 17 astronauts to travel farther than previous missions (22 miles in total). (Courtesy of NASA.)

Los Alamos National Laboratory (LANL) began as part of the World War II Manhattan Project's centralized facility for nuclear scientific research. Prior to this, atomic research was spread around the country at many universities. Scientific director J. Robert Oppenheimer spent many years of his youth in New Mexico and recommended the Los Alamos Mesa to Gen. Leslie Groves. The project's early days brought thousands of workers to this remote New Mexico town. (Courtesy of Los Alamos National Laboratory.)

A display at the Bradbury Science Museum shows a Vela Hotel satellite and B83 bomb casing. On September 22, 1979, a double flash of light was detected near the Prince Edward Islands by an American Vela Hotel nuclear monitoring satellite. There is uncertainty as to the true nature of the incident, though it is widely speculated to have been the result of a nuclear detonation, completely surprising the United States. (Courtesy of Los Alamos National Laboratory.)

A close-up picture shows the Curiosity rover's ChemCam, on display at the Bradbury Science Museum. The instrument uses a laser to vaporize Martian rock and soil material, illuminating the material in a bright flash, allowing three onboard spectrometers to analyze the elemental composition. ChemCam and the rover's plutonium power source were produced at LANL. (Courtesy of Los Alamos National Laboratory.)

A model of the FORTE (Fast On-orbit Rapid Recording of Transient Events) satellite resides in the Bradbury Science Museum at LANL. The satellite and sensors were developed jointly by LANL and Sandia National Laboratories in Albuquerque as a test bed for technologies applicable to US nuclear detonation detection systems. Monitoring compliance with arms-control treaties prevents politically sensitive nuclear events like 1979's "Vela Incident" from reoccurring or escalating. (Courtesy of Los Alamos National Laboratory.)

A model of the Genesis spacecraft's collector resides in the Bradbury Science Museum. Launched in 2001, Genesis captured samples of solar wind particles by focusing isotopes of light elements onto a small target and concentrating them, allowing precise measurements of isotopic abundances. The spacecraft and samples crash-landed back on Earth in 2004; however, several of the sample collectors were undamaged, providing scientists with their first close look at solar wind. (Courtesy of Los Alamos National Laboratory.)

The Boy Scouts of America's Philmont Scout Ranch hosted NASA astronauts during field exercises. Aside from operating their spacecraft and surviving extravehicular activity (spacewalks), astronauts were also trained to be scientific observers for Earth-bound scientists. The geologic field trips in northern New Mexico also proved useful for future moonwalkers, some of whom stand here in a group photograph. All of the astronauts in white pants are Eagle Scouts. (Courtesy of NASA.)

Astronauts Dave Scott (left), Neil Armstrong (second from left), and Roger Chaffee (second from right) discuss gravity meter data with Joel Watkins (right) of the US Geological Survey. Armstrong would command Apollo 11 and become the first human on the surface of the moon; Scott would command Apollo 15, the first mission with the Lunar Rover. Sadly, Chaffee was killed during the disastrous Apollo 1 fire with Gus Grissom and Ed White in 1967. (Courtesy of US Geological Survey.)

Boy Scouts of America (BSA) executive Ray Bryan (right) talks with astronauts Ed White (center), Alan Bean (behind White), and Neil Armstrong (left) after they were presented with jac-shirts with black bull logos. The connection between BSA and the space program was strong and continues to the present day. Of the 24 men to travel to the moon, 21 were Scouts. (Courtesy of Philmont Museum–Seton Memorial Library.)

A group of astronauts expresses thanks to BSA executive Ray Bryan for their Philmont coats. It is unknown how many of the astronauts, present for the jac-shirt ceremony, had been to Philmont before. Whether experiencing a trek, participating in the training center, or being one of the Philmont staff, the wearer of the jac-shirt stands apart as coming to the "Land of Enchantment" for a life-changing experience with the BSA. (Courtesy of Philmont Museum–Seton Memorial Library.)

Future moonwalkers Eugene Cernan (left) and Charles "Pete" Conrad (right) locate their position at Philmont Scout Ranch. Conrad would later command Gemini 11 and become the third man to walk on the moon on Apollo 12. Cernan flew on Gemini 9-A and went to the moon twice, once on Apollo 10, coming within eight miles of the lunar surface, and later becoming the last man to walk on the moon on Apollo 17. (Courtesy of Philmont Museum–Seton Memorial Library.)

US Geological Survey geologist Al Chidester (center) discusses surface texture with astronauts Walt Cunningham (left) and Neil Armstrong (right). Cunningham would fly on Apollo 7 as the Lunar Module pilot, even though the mission didn't have a Lunar Module. Armstrong would become the first human to walk on the moon; his two-and-a-half-hour moonwalk collected almost 50 pounds of lunar sample material. (Courtesy of Philmont Museum–Seton Memorial Library.)

Astronauts Russ Schweickart (standing) and Pete Conrad (right) attempt to locate their position by resection while geologist Al Chidester (center) looks on. Resection is the method of locating one's position on a map by determining grid azimuth to at least two well-defined locations that can be pinpointed on the map. Philmont contains many such landmarks, so their task may not have been as difficult as it would have been on the lunar surface. (Courtesy of Philmont Museum–Seton Memorial Library.)

Astronauts Jim Lovell (right, standing) and Ted Freeman (left, standing) discuss measuring a section with a Jacob's staff with geologist Uel Clanton at Philmont in June 1964. Sadly, Freeman would be the astronaut corps' first fatality four months after this photograph was taken. Jim Lovell would become the space program's most experienced astronaut, with four spaceflights (Gemini 7 and 12 and Apollo 8 and 13) and over 715 hours in space. (Courtesy of Philmont Museum–Seton Memorial Library.)

This wide-angle shot was taken from the edge of the Valles caldera. Apollo astronauts were brought to the caldera in 1964 and 1966 for geologic field training. Volcanologist Bob Smith trained them on what lunar samples to look for and what to bring back. Additionally, Smith was one of the first scientists to examine the moon rock samples when they arrived on Earth. (Courtesy of Valles Caldera National Preserve.)

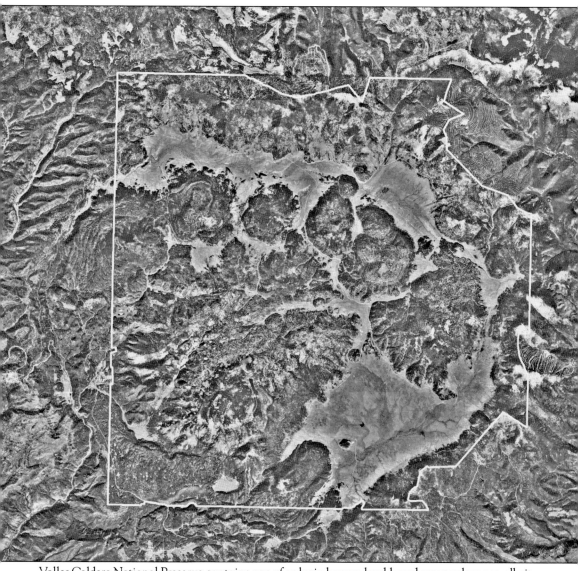

Valles Caldera National Preserve contains one of only six known land-based super volcanoes, albeit the smallest one of the six. The circular rim of the caldera measures 12 miles in diameter, and it is the most thoroughly studied caldera complex in the United States. The map's white border denotes the size of the national preserve complex. (Courtesy of Valles Caldera National Preserve.)

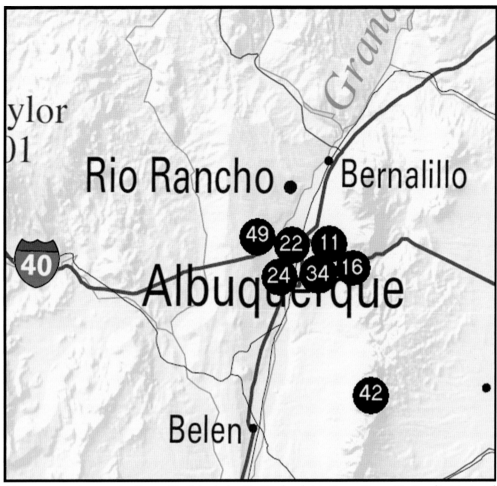

Albuquerque sites on the Space Trail include: 11. Kirtland Air Force Base; 16. Sandia National Laboratories; 22. University of New Mexico Campus Observatory; 24. Lovelace Medical Center; 34. National Museum of Nuclear Science and History; 42. University of New Mexico's Capilla Peak Observatory; and 49. New Mexico Museum of Natural History & Science and Planetarium.

Three

ALBUQUERQUE

Founded in 1706 as the Spanish colonial outpost of Ranchos de Albuquerque, the city has held importance in cultural, farming, and strategic military areas since its early days. Around the beginning of the 20th century, the population began to grow, and amenities such as railways and secondary education began to arrive. The area's dry climate began to attract tuberculosis patients searching for a cure in the early 20th century. Dr. William Randolph Lovelace contracted tuberculosis while earning his medical degree and moved his family (including infant son William II) to New Mexico. The son would famously help certify the first Mercury astronauts for flight in 1960 and also run a similar unofficial program for the "Mercury 13," the first female astronaut candidates.

The story of Kirtland Air Force Base merges the lineage of three bases (Kirtland, Manzano, and Sandia) combined in 1971 from World War II–era and early Cold War infrastructure around Albuquerque. The Air Force Research Laboratory at Kirtland performs research in many areas of high altitude and space-related activities. The facilities at Sandia National Laboratories also push the limits on energetic materials research and radiation effects on satellites. Sandia advisors were instrumental during the investigation into the 2003 space shuttle *Columbia* accident.

The Department of Physics and Astronomy at the University of New Mexico has observatories on campus and on the nearby Manzano Mountains at Capilla Peak. UNM's research covers the span of astronomical topics, from cosmology to cosmic radiation and radio astronomy to supernova physics. UNM's astronomy is truly world-class, and the department has collaborated with New Mexico Tech and New Mexico State University as well as the US Air Force at Kirtland.

The final noteworthy stop in Albuquerque has actually had three locations and three names during its existence. In 2009, the National Museum of Nuclear Science and History finally put down roots on the eastern edge of the Kirtland Air Force Base boundaries. Within its walls, the race to harness the nuclear genie is chronicled. Outside, many nuclear-related artifacts are splayed across the grounds, including military aircraft that participated in actual live nuclear tests during the 1950s and 1960s.

Kirtland Air Force Base's entry sign on Gibson Boulevard is pictured here. The base is located in the southeast corner of Albuquerque and occupies over 52,000 acres. Originally called Albuquerque Army Air Field, it was renamed in 1942 after Col. Roy C. Kirtland, one of the Army's first pilots trained by the Wright brothers. After the existence of the atomic bomb was revealed, Kirtland's focus changed to supporting the US nuclear research program.

The Starfire Optical Range (SOR), located in the Manzano Mountains behind Kirtland Air Force Base, contains a 3.5-meter telescope with adaptive optics, a 1.5-meter telescope, and a 1-meter beam director. The technologies tested at SOR help refine the Air Force's space situational awareness sensors around the globe. As satellites get smaller and the number of space objects increases, advances in imaging and identification of space objects are paramount. (Courtesy of the US Air Force.)

The Air Force Nuclear Weapons Center (AFNWC) was created in 2006 to ensure safe, secure, and reliable nuclear weapon systems and support to the national command structure. Embarrassing incidents such as the mistaken transfer of live nuclear warheads in 2007, accidental shipment of re-entry vehicle parts to a close ally, and missile crew procedural violations regarding nuclear surety rules reinvigorated national attention on the nuclear weapons enterprise. (Courtesy of the Air Force Historical Research Agency.)

The SpaceLoft-6 sounding rocket launched seven payloads on April 5, 2012, from Spaceport America. The Operationally Responsive Space (ORS) office at Kirtland aims to have the ability to go from payload call-up to launch and operations within a matter of days. SpaceLoft-6 flew for 13 minutes to an orbital altitude of 70 miles, landing 33 miles away from its launch pad, within the confines of White Sands Missile Range. (Courtesy of the US Air Force.)

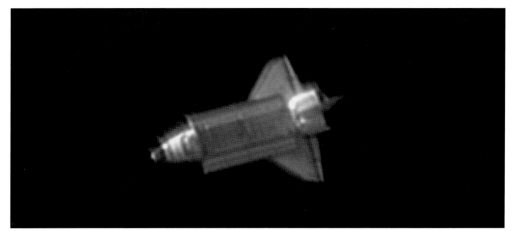

An optically enhanced image from the Air Force Maui Optical and Supercomputing (AMOS) observatory shows the space shuttle *Atlantis* during its last mission. The Air Force Research Laboratory's Directed Energy Directorate drives the optical research at AMOS, Starfire Optical Range, and North Oscura Peak on White Sands Missile Range for the mission requirements of worldwide space situational awareness. (Courtesy of the US Air Force.)

When originally approached about Felix Baumgartner's record-breaking jump plans in 2007, Kirtland officials declined assistance, not believing the activity had a science/technology perspective. They reversed their decision 18 months later and provided assistance and personnel such as Air Force Research Laboratory (AFRL) staff members with over 20 years of balloon-launching expertise. On October 14, 2012, Baumgartner made the highest jump ever recorded at 128,100 feet, freefalling at 834 miles per hour over the southeastern New Mexico skies. (Courtesy of Red Bull Stratos.)

Around 2007, an AFRL-sponsored payload team poses before a Minotaur 1 launch vehicle containing the TacSat-2 satellite. From left to right are 1st Lt. Joshua Johnson, Capt. Khirah Morgan, Capt. Hiro Ababon, Maj. Robert Douglass, and Col. Sam McCraw, all from Kirtland's Space Development and Test Wing. (Courtesy of the US Air Force.)

The University of New Mexico's Campus Observatory contains a 14-inch telescope with a CCD camera inside its protective dome. The observatory is open every Friday evening during UNM's fall and spring semesters for the locals to stargaze. Volunteers from The Albuquerque Astronomical Society (TAAS) and UNM's Department of Physics and Astronomy assist with interpretation and viewing through additional telescopes. When not stargazing, the observatory assists in faculty research in areas such as cosmology, galactic astronomy, and solar physics. (Courtesy of Jon Lewis, UNM Department of Physics and Astronomy.)

This shot of the Astronomical LIDAR for Extinction equipment at the UNM Campus Observatory shows the LIDAR (light detection and ranging) beam shooting into space. The project will precisely measure the earth's atmosphere, allowing corrections for the significant distortion effects of the atmosphere on astronomical observations. (Courtesy of Jon Lewis, UNM Department of Physics and Astronomy.)

Sandia National Laboratories began as an offshoot of the Los Alamos National Laboratory's Z-Division to handle future weapons development and testing. The facilities at Oxnard Field were assigned to the Armed Forces Special Weapons Project and given the name Sandia Base after the nearby Sandia Mountains. On July 1, 1971, Sandia Base was merged, along with Manzano Base (a weapons stockpile site), into Kirtland Air Force Base. (Courtesy of Sandia National Laboratories.)

The National Solar Thermal Test Facility is an eight-acre field of 220 solar-collection heliostats aimed at a 200-foot-tall tower. Material samples are installed in the tower's wind tunnel and exposed to solar beams approximately 1,500 times the intensity of the sun on a clear day. This image shows a material sample being heated to simulate travel through the atmosphere of Titan, Saturn's largest moon. (Photograph by Randy Montoya, courtesy of Sandia National Laboratories.)

Kenneth Gwinn (right) points to an image of a Sandia computer model/ analysis of foam impacting the leading edge of the space shuttle to David Crawford (left) and Kurt Metzinger (sitting). In addition to the *Columbia* disaster, Sandia personnel have been instrumental in assisting investigations into the 1989 turret explosion aboard the USS *Iowa*, the 1996 TWA Flight 800 accident, and the Unabomber domestic terrorism case. (Photograph by Randy Montoya, courtesy of Sandia National Laboratories.)

After *Columbia*'s heat shield failed in 2003, causing the tragic accident that took the lives of all seven on board, Sandia developed the Laser Dynamic Range Imager (LDRI), which generates three-dimensional images from two-dimensional video. The LDRI Orbiter Inspection System (LOIS), attached to the orbiter's robotic arm, scans the heat shield twice, once 18 hours after liftoff and then again the day before re-entry, to ensure that the heat shield is intact. (Photograph by Randy Montoya, courtesy of Sandia National Laboratories.)

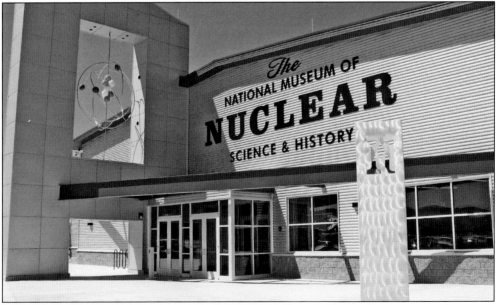

The mission of the National Museum of Nuclear Science & History is to serve as America's resource for nuclear history and science. The museum tells the story of the Atomic Age from early research of nuclear development through today's peaceful uses of the technology. The museum is the only Smithsonian affiliate in the city of Albuquerque and was chartered by Congress in 1991. (Courtesy of National Museum of Nuclear Science & History.)

One of the museum's missions is to educate, with many interactive displays and a science summer camp program with 300 day-campers learning about ecology, robotics, flight, engineering, medicine, and general science. (Courtesy of National Museum of Nuclear Science & History.)

A Redstone rocket sits outside of the National Museum of Nuclear Science & History in Albuquerque, New Mexico. The museum's five-acre outdoor Heritage Park is replete with nuclear artifacts—nuclear delivery aircraft, rockets, missiles, and even a nuclear submarine sail. The Redstone rocket is one of three in the state of New Mexico; the others reside at White Sands Missile Range and in the NMMSH's storage lot in Alamogordo. (Courtesy of NMMSH.)

The backbone of America's nuclear retaliation force for over 50 years, a B-29 Superfortress and B-52 Stratofortress sit inside the National Museum of Nuclear Science & History's storage lot. The B-52B, serial number 52-0013, is one of a few to have actually dropped a nuclear weapon. It was assigned to Operation Redwing and dropped the Cherokee device on May 21, 1956, resulting in a 3.8-megaton explosion. (Courtesy of the National Museum of Nuclear Science & History.)

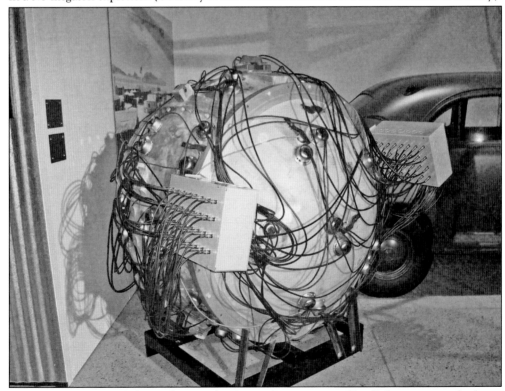

The unwieldy design of "the Gadget," as the Trinity device was unofficially known, is on display at the National Museum of Nuclear Science & History. The wires and nodes on the spherical container allowed explosive packages to compress the internal plutonium sphere to critical mass, thereby initiating an uncontrolled fission reaction. The design was only for testing and was simplified for weaponized designs. (Courtesy of the National Museum of Nuclear Science & History.)

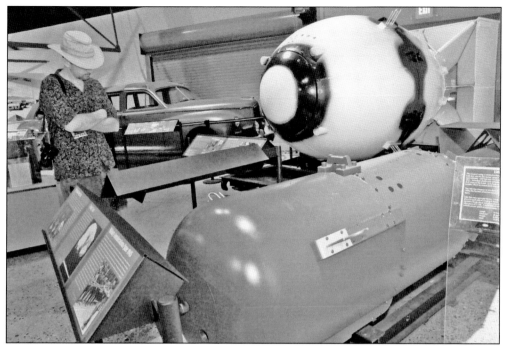

A museum patron inspects bomb casings of Fat Man and Little Boy, the first two atomic weapon designs. Fat Man, similar to the Trinity test "Gadget," used an implosion-type design, symmetrically crushing a plutonium sphere to critical mass and a huge explosion. Little Boy was a gun-type design, firing a hollow "bullet" onto a cylindrical target to create critical mass. Gun-type designs have been all but abandoned in favor of implosion types. (Courtesy of the National Museum of Nuclear Science & History.)

The lobby of the museum prominently displays the Smithsonian affiliate flag. Being a member of the affiliate program has allowed the museum to borrow and display a wide array of artifacts relating to nuclear science and history. While the museum's central topic is heavily inundated with warfare, the museum offers a balanced view of nuclear science and peaceful means, such as the Pioneers of the Atom display.

Lovelace Medical Center in Albuquerque, New Mexico, continues to support the military today with its facility adjacent to Kirtland Air Force Base. Dr. William Randolph Lovelace II was a military flight surgeon, studying the problems of high-altitude flight in the 1930s at Wright Field, Ohio. As head of NASA's Life Sciences, he played a critical role in the selection of the seven Mercury astronauts. In 1964, he was appointed NASA's director of space medicine. (Courtesy of NMMSH.)

Geraldyn "Jerrie" Cobb poses by a mockup of the Mercury spacecraft. In 1960, Jerrie Cobb was selected by Dr. Lovelace to undergo medical testing identical to the tests Mercury astronauts underwent for selection. Cobb was the first female to successfully pass all three phases of testing. Cobb and Dr. Lovelace recruited more aviatrixes for testing, narrowing down the field to 13 before the government halted the unsanctioned program. (Courtesy of NASA.)

A full-scale replica of the Mars Exploration Rovers, also known as Spirit and Opportunity, is on display at the New Mexico Museum of Natural Science & History. Sandia National Laboratories has been a pioneer in the use of robotic systems for tasks ranging from interplanetary exploration to the security of the nation's nuclear arsenal. (Courtesy of NMMSH.)

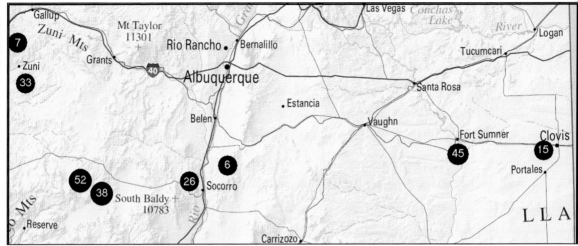

Space Trail sites in central New Mexico include: 6. Scholle Crab Nebula Supernova Petroglyph; 7. Zuni Pueblo Crab Nebula Supernova Petroglyph; 15. Cannon Air Force Base; 26. New Mexico Institute of Mining and Technology (New Mexico Tech); 33. Zuni Salt Lake; 38. Very Large Array Radio Telescope; 45. Columbia Scientific Balloon Facility; and 52. Magdalena Ridge Observatory.

Four

CENTRAL

The central part of New Mexico comprises a small strip running the width of the state from south of Albuquerque to just north of Las Cruces. Within this part of the state, active astronomical research is being performed constantly, with balloons launched from Fort Sumner's Columbia Scientific Balloon Facility (CSBF), frequent sky surveys from the Karl G. Jansky Very Large Array (VLA), and the interferometry techniques used at the Magdalena Ridge Observatory.

The CSBF, established in 1961 as the National Scientific Balloon Facility, is a NASA facility responsible for providing launch, tracking and control, airspace coordination, telemetry and command systems, and recovery services for unmanned, high-altitude balloons.

The cutting edge of radio astronomy resides at the campus of the New Mexico Institute of Mining and Technology, colloquially known as New Mexico Tech, home of the National Radio Astronomy Observatory's Array Operations Center. Coordinating the actions of the VLA and the worldwide Very Long Baseline Array, the radio antennas work together as a giant array, simulating an antenna roughly 22 miles across—a big "ear" to hear faint intergalactic sounds.

The Scholle Petroglyph site and Zuni Pueblo petroglyphs display drawings of the supernova explosion of 1054 AD and illustrate other ancient observation sites within the state. These areas are considered sacred and are off-limits to outsiders, protecting the drawings from unnecessary damage by visitors.

The final location within the central zone is Cannon Air Force Base, home to the 27th Special Operations Wing's air commandos. As part of a long lineage of special operations, the wing carries a tradition of performing unconventional missions with men and women of uncommon bravery. Historical predecessors of the air commandos performed retrieval of space capsules; it is presumed the commandos still practice the clandestine techniques involved in grabbing important cargo, human or otherwise.

Just as the Navajo have incorporated their creation myths into pictorial drawings, the Zuni people created drawings of the supernova of 1054 AD. The explosion created the Crab Nebula, as seen here. (Courtesy of New Mexico Tech/Magdalena Ridge Observatory.)

The Magdalena Ridge Observatory (MRO) is managed by New Mexico Tech in Socorro. The observatory consists of two major facilities: a fast-tracking 2.4-meter telescope and an infrared/optical interferometer. The telescope has high slewing and tracking rates, making it a superb instrument for the study of fast-moving objects and targets of opportunity. NASA contributes 30 percent of the telescope funding for near-earth object follow-on studies. (Courtesy of New Mexico Tech/ Magdalena Ridge Observatory.)

The 2.4-meter MRO Scientific Instrument for NEO, GEO, and LEO Exploration (SINGLE) Telescope can slew its view 10 degrees per second, enabling it to observe objects in low-earth orbit (LEO). It accomplished its first viewing activities, also known as "first light," on October 31, 2006. It began regular operations two years later; one of its first major tasks was tracking asteroid 2007 WD5 for NASA. (Courtesy of New Mexico Tech/Magdalena Ridge Observatory.)

The telescope's primary mirror was originally a backup mirror for the Hubble Space Telescope. As of October 2008, the facility is assisting NASA in the tracking of LEO objects. MRO was one of three New Mexico–based telescopes (Tortugas Mountain in Las Cruces and the New Mexico Tech Etscorn Campus Observatory) assisting in the LCROSS Project, observing controlled impacts on the southern polar region of the moon. (Courtesy of NM Tech/Magdalena Ridge Observatory.)

Construction contracts from the US Department of Transportation have provided for the planned Magdalena Ridge Observatory visitors center by creating an access road and maintenance facility on the mountain. MRO has an elevation of over 10,600 feet above sea level, and it features an array of optical and infrared telescopes used primarily for astronomical research. The creation of visitor facilities would greatly increase tourism in the Socorro area. (Courtesy of New Mexico Tech/Magdalena Ridge Observatory.)

A view down the MRO optical interferometer shows the great length of the combining path. The beam combining and delay line facilities are routed along the 200-meter-long portion of the hilltop building. Beams enter the building from the center of the Y-array from the arms containing the telescopes. First light for the interferometer is expected in 2015. (Courtesy of New Mexico Tech/Magdalena Ridge Observatory.)

The older logo of the National Radio Astronomy Observatory (NRAO) prominently displays a radio telescope, the organization's instrument of choice. NRAO is an organization funded by the National Science Foundation with multiple locations across the United States. The most public display of NRAO activities is near Socorro, New Mexico, at the Very Large Array and Green Bank, West Virginia. (Courtesy of NRAO/National Science Foundation.)

The Pete V. Domenici Science Operations Center, headquartered on the campus of New Mexico Tech, doubles as the Array Operations Center (AOC) for the VLA and the Very Large Baseline Array (VLBA). From inside the AOC, operators can remotely control the 10 VLBA telescopes over a network connection, in addition to monitoring the health and status of every antenna and receiving its data downlinks. (Courtesy of NRAO/National Science Foundation.)

Very Large Array dishes dot the Plains of San Agustin in west central New Mexico. On March 31, 2012, the array was renamed the Karl G. Jansky Very Large Array after the discoverer of radio waves emanating from the Milky Way galaxy in 1931. Karl Jansky subsequently became one of the founding fathers of radio astronomy. Recent upgrades to the array replaced 1970s-era electronics with state-of-the-art receiving and processing equipment. (Courtesy of NRAO/ National Science Foundation.)

Similar to a camera lens's zoom feature, the VLA has four different configurations, labeled A through D, each corresponding to the separation length. In the A configuration, the telescopes are extended over the 13-mile distance of each arm. This simulates a single radio antenna that is 22 miles in diameter. In contrast, the D configuration has the arms only .4 miles from the center. (Courtesy of NRAO/National Science Foundation.)

In a feat of deft project management, the VLA scheduling includes having one dish in depot maintenance at any time. The dish is brought along the railroad tracks and placed inside the building. Engineers and workers then perform routine maintenance, as well as any equipment upgrades, such as the change to VLBA configuration standards. (Courtesy of NRAO/National Science Foundation.)

The author stands near the edge of a VLA dish undergoing maintenance. The clear spaces of the Plains of San Agustin can be seen over the dish's edge. Unlike the US National Radio Quiet Zone established in West Virginia to limit radio interference on astronomical research, the mountains and sparse population in the west central part of New Mexico act as a natural buffer, with no legislation needed. (Courtesy of the Clark family.)

Kevin Clark (left), Sara Clark (center), and the author (right) stand next to the feed horn containing sensor instruments in the dish's center. As a comparison, the author stands roughly six feet tall—the large size of the sensors corresponds to the receiver dish's listening frequencies. Below the feed horn is the vertex room, which houses additional equipment used to process and send the received information. (Photo courtesy of the Clark family.)

The author stands in front of one of the VLA's 25-meter dishes. When array reconfigurations or repairs are necessary, the 230-ton dish is lifted off its four concrete pedestals by a mechanized transporter (seen at base of dish) and transported along railroad tracks. The transporter, dubbed "Hein's Trein," was conceived by Dr. Hein Hvatum, who was responsible for the VLA's construction in the 1970s and early 1980s. (Courtesy of the Clark family.)

Aside from its astronomical duties, the VLA has been used as a location for a few Hollywood blockbuster movies. In the 1997 movie adaptation of Carl Sagan's book *Contact*, the VLA is fictionally portrayed as a part of the Search for Extra Terrestrial Intelligence (SETI) project. In the science-fiction film *Terminator: Salvation*, the array is part of the Skynet artificial intelligence network and is destroyed by resistance fighters. (Courtesy of NRAO/National Science Foundation.)

The VLA observes the sky 24 hours a day in all but the most severe weather conditions. The telescope can currently observe 10 different frequency bands ranging from approximately 74 megahertz (four-band) all the way up to 50 gigahertz (Q-band), and in a given day may observe astronomical objects such as supernovae, hydrogen gas clouds, gamma ray bursts, active galactic nuclei (AGNs), and occasionally even the sun or planets. (Courtesy of NRAO/National Science Foundation.)

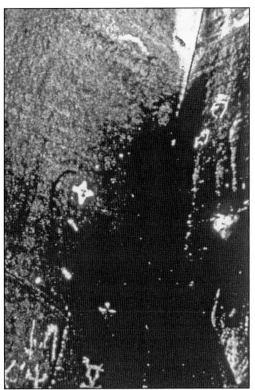

The Scholle petroglyph purports to show the creation of the Crab Nebula in 1054 AD. Unlike many Southwest petroglyphs, the Scholle image was dated years after the event. The crescent and star were painted with an emulsion of aluminum powder and water for increased contrast. Other images at Scholle show the Morning Star pictogram, complete with claws, demonstrating its function as a war captain in local mythology. (Courtesy of NMMSH.)

A balloon is inflated at the Columbia Scientific Balloon Facility (CSBF), near Fort Sumner. The facility has been used by NASA since 1987, when it was established as the National Scientific Balloon Facility. The facility's Texas headquarters was right under the landing path of the ill-fated STS-107 mission in 2003. All associated facilities were renamed in February 2006 to honor the final crew of the space shuttle *Columbia*. (Courtesy of the Columbia Scientific Balloon Facility.)

CSBF is managed by NMSU's Physical Science Laboratory and is administered by Goddard Space Flight Center's Wallops Flight Facility Balloon Program Office. This image shows the Sunrise project with its launching group. Sunrise is a balloon-borne solar telescope able to give clearer pictures of the sun than ground-based instruments. On its inaugural test flight from CSBF in 2007, it performed flawlessly, and was later found in a Texas farmer's field. (Courtesy of the Columbia Scientific Balloon Facility.)

The Wallops Arc Second Pointer (WASP) platform is readied for flight by the Mobile Launch Vehicle (MLV), a one-of-a-kind transporter truck. The WASP gondola allows very controlled pointing for scientific payloads while airborne. Its maiden flight took place on October 7, 2011, and successfully held the balloon at an altitude of 125,000 feet. Solar studies on high-altitude balloons take place in Antarctica, with payloads aloft for up to 50 days. (Courtesy of the Columbia Scientific Balloon Facility.)

A balloon mission launches from the CSBF. While considered anachronistic by some research scientists favoring satellites and space launches, high-altitude research still has much to offer. The military's resurgent interest in high-altitude balloons at "near space" altitudes (90,000 to 135,000 feet) for communications is just one area that will blossom in future decades. (Courtesy of the Columbia Scientific Balloon Facility.)

Outside of Clovis, New Mexico, resides the front gate of Cannon Air Force Base. Originally established as Portair Field in the 1920s, the base became a World War II–era training field for the Army Air Forces. Fundamentally a fighter jet base for over 50 years, the location was transferred to Air Force Special Operations Command in 2006 and charged with training air commandos in a variety of special operations missions. (Courtesy of the 27th Special Operations Wing Public Affairs Office.)

An MQ-9 Reaper unmanned aerial system sits on Cannon Air Force Base's tarmac. The strategic shift from manned combat aircraft to unmanned systems provided a needed boon to New Mexico's economy. The control of aircraft by ground station through satellite link creates a critical path for operators. Fortuitously, the state contains many areas free of radio frequency interference and extremely clear views of the southern night sky for mostly accident-free flying. (Courtesy of the 27th Special Operations Wing Public Affairs Office.)

A 6594th Recovery Control Group JC-130 Hercules approaches an airborne film bucket, recently returned from space. The missions and equipment of Cannon Air Force Base's present-day special operations warriors are historically and spiritually connected to these legacy missions, with similar capabilities to grab clandestine packages, human or otherwise, and return them safely home. As part of Air Force Special Operations Command's "quiet professionals," there is little public acknowledgement of their missions. (Courtesy of the 6594th Test Group Web Site.)

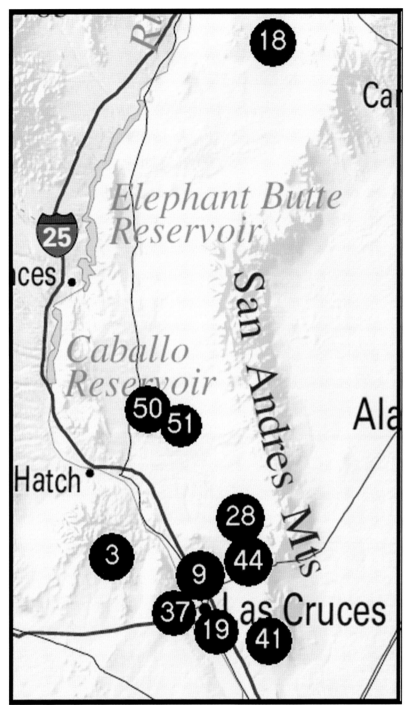

Southwest sites on the Space Trail include: 3. Tenuco Observatory; 9. New Mexico State University; 18. Trinity Site; 19. NMSU Physical Science Laboratory; 28. White Sands Test Facility; 37. Clyde W. Tombaugh Campus Observatory; 41. Tortugas Mountain Planetary Observatory; 44. Tracking and Data Relay Satellite System; 50. Spaceport America; and 51. Virgin Galactic International Headquarters.

Five

SOUTHWEST

The storied lines of the past, present, and future intersect at the southwestern quadrant of New Mexico. Inspired by Francisco Vásquez de Coronado's valiant search for the Seven Cities of Gold, the conquistadores' arrival to the southwest was later followed by Juan de Oñate's sacking of the New Mexican landscape and population. The path along the Rio Grande fortified the use of the area as a travel stop along the El Camino Real de Tierra Adentro, a trade route running from Mexico City to Santa Fe. Decades later, the trade route's modern-day successor, Interstate 25, would be nicknamed "America's Nuclear Highway," running along nuclear weapons testing, development, and deployment sites.

The largest city in the southwest quadrant is Las Cruces, Spanish for "The Crosses." Las Cruces is home to New Mexico State University (NMSU), the state's only land grant university and home of the New Mexico Space Grant Consortium. Within the halls of NMSU, astronomical wonders are researched, engineering marvels created, and prototypes tested to further mankind's celestial knowledge.

On the outskirts of Las Cruces lies the White Sands Test Facility (WSTF), a satellite facility to the Johnson Space Center. The facility was created in 1962 to test propulsion systems in the Apollo program. Lunar Module landing engines, launch escape systems, the Viking Mars landers, and the space shuttle are just a few of the programs supported by WSTF. On the same property reside the ground terminals for the Tracking and Data Relay Satellite System, or TDRSS. The backbone to all US space communications, the jewel that is TDRSS has kept NASA at the scientific forefront by controlling many high-priority space missions simultaneously, routing critical data back to the mission owners.

The future of commercial spaceflight resides in the small town of Upham, home of Spaceport America. Run by the state of New Mexico, the spaceport houses Virgin Galactic, the world's first commercial spaceflight company. Already a trailblazer for its forwarding of space tourism, Spaceport America expects to see the first continuing commercial spaceflights in 2014, but time will tell.

Hadley Hall at NMSU forms the centerpiece of the campus. Originally opened as the Las Cruces College in 1888, it went through two more name changes until settling on NMSU in 1960. The university is home to the New Mexico Space Grant Consortium, whose goal is to establish and maintain a national network of universities with interests and capabilities in aeronautics, space, and related fields (Courtesy of NMSU University Communications.)

The main observatories at NMSU are named after Clyde Tombaugh, discoverer of the dwarf planet Pluto in 1930. Tombaugh worked at White Sands Missile Range in the early 1950s and taught astronomy at NMSU from 1955 until his retirement in 1973. He died on January 17, 1997, in Las Cruces. A small portion of his ashes is being carried aboard the New Horizons space probe, expected to reach Pluto in July 2015. (Courtesy of NMSU University Communications.)

Nancy Chanover, astronomy student and Tombaugh Scholar, poses in one of the three main NMSU campus observatory domes around 2000. The observatories are an integral part of the campus astronomy department. Undergraduate astronomy students are required to obtain three observations over the course of their 16-week class for analysis. Now a doctor, Chanover's research involves the study of planetary atmospheres using visible and infrared imaging and spectroscopic techniques. (Courtesy of NMSU University Communications.)

Two of the three domes of the Tombaugh Observatory are pictured. The astronomy department holds open house events one Friday night a month for the public to use the equipment. They occur near the time of the first-quarter moon each month from September through May and include an evening program with a short presentation before guided observing through telescopes.

The Tortugas Mountain Observatory is positioned on Tortugas Mountain, otherwise known as "A Mountain" due to a large white letter denoting NMSU's Aggies nickname. Used regularly from 1963 through 2000, the building and equipment are receiving increased attention by NMSU astronomy professors and students and NASA in the last decade. Light pollution within the Mesilla Valley is minimal, so the mountain's altitude provides pristine views for stargazing. (Courtesy of NMSU Department of Astronomy.)

NMSU Department of Astronomy head Dr. Jon Holtzman poses inside the refurbished dome at Tortugas Mountain Observatory around 2008. Virtually abandoned at the beginning of the 21st century, the inside of Tortugas Mountain Observatory looked like an archaeological excavation site before refurbishment. Work centers with decades-old books and outdated computer equipment lined the countertops. The site fell into disuse around 2000, and it became a project for refurbishment in 2008. (Courtesy of NMSU Department of Astronomy.)

While Space Murals Inc. is not officially part of the Space Trail, it has been an excellent resource for space aficionados and tourists. Pieces of space shuttle heat tiles, mock-ups of the International Space Station, static displays of V-2 wreckage, and Army missiles line the property. The site is located at the end of NASA Road, the street that leads to the White Sands Test Facility and TDRSS Ground Sites.

Workmen put the finishing touches on the sign for the Apollo Site, as White Sands Test Facility was originally known from 1962 until 1965. The first rocket engines were tested at the site in 1964. Apollo's propulsion systems were critical to the program—first allowing the astronauts to land safely on the moon and then allowing them to depart the lunar surface to return home to Earth. (Courtesy of the White Sands Test Facility Public Affairs Office.)

A mock-up of the Apollo Command and Service Modules resides on WSTF property. The display is a curt reminder to engineers that their work forwarding national prestige was important, but so were the lives of the astronauts aboard the spacecraft. In a macabre piece of space history, President Nixon penned an alternate speech after Apollo 11's landing on the moon, to be used if the astronauts could not take off from the lunar surface. Thanks in large part to WSTF, that eventuality did not occur. (Courtesy of the White Sands Test Facility Public Affairs Office.)

The White Sands Test Facility was originally established to provide a central test location for Apollo program propulsion systems and integrated testing. The 88-square-mile portion of White Sands Missile Range was selected in 1962. The site performs propulsion tests in static test stands and live fires on the nearby missile range as the official Johnson Space Center (JSC) Propulsion Systems Development Facility. (Courtesy of the White Sands Test Facility Public Affairs Office.)

An engineering boilerplate of the Apollo Lunar Module resides at WSTF as a holdover from testing in the 1960s. Large cylindrical and spherical pressure tanks are shown here, containers for oxidizer and fuel during the lunar stays. Unlike other Lunar Module static displays around the world, WSTF's model was never built as a full craft—it existed solely as a propulsion test model and did not receive a Lunar Module serial number. (Courtesy of the White Sands Test Facility Public Affairs Office.)

A December 14, 1972, photograph shows the ascent stage of the Apollo 17 Lunar Model leaving the lunar surface. The ungainly design of the Lunar Module is clear in the photograph, with the triangular windows for the two astronauts, the rectangular extravehicular-activity hatch visible in the front, and other features normally obscured in other photographs. (Courtesy of NASA.)

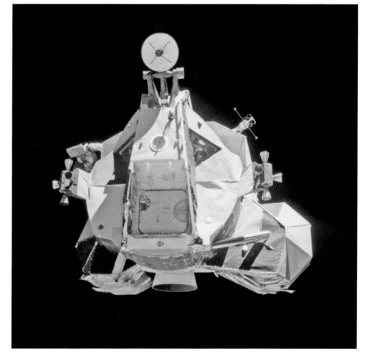

This interior view of the Lunar Module cockpit shows the limited real estate. Built by Grumman in Bethpage, New York, the Lunar Module was considered the world's first true spacecraft, since its design did not follow aerodynamic principles and was meant to be used only in space. The craft was also statistically the most reliable component of the Apollo program, with no vehicle suffering a mission-impacting failure. (Courtesy of Tyler Rubach.)

The skeletal remnants of a Lunar Module test article reside at WSTF. The model is not cleared for public viewing, due to the presence of poisonous chemicals inside the tanks and interior surfaces from its testing program. The photographs show how delicate and small a craft the LM actually was, being home to two astronauts for up to 72-hour stays on the moon. (Courtesy of the White Sands Test Facility Public Affairs Office.)

Astronauts Deke Slayton (left, standing), L. Gordon Cooper (center), and Alan Shepard (right, seated) are shown a model of the Apollo Launch Escape System (LES) by Maj. Gen. J. Frederick Thorlin, White Sands Missile Range (WSMR) commanding general (left, seated), and W.E. Messing, NASA resident manager (right, standing). On May 13, 1964, the astronauts witnessed the Apollo Little Joe II launch at WSMR's Launch Complex 36. (Courtesy of the White Sands Test Facility Public Affairs Office.)

WSTF is still a critical part of the nation's space program. The positioning of the site allows the military and civil programs to use governmental resources (i.e., WSMR) to test, and also allows commercial spaceflight at Spaceport America to capitalize on WSTF's propulsion expertise. If the concept of single-stage-to-orbit reaches operational maturation, its success will be due to the bedrock expertise provided by NASA's WSTF employees. (Courtesy of the White Sands Test Facility Public Affairs Office.)

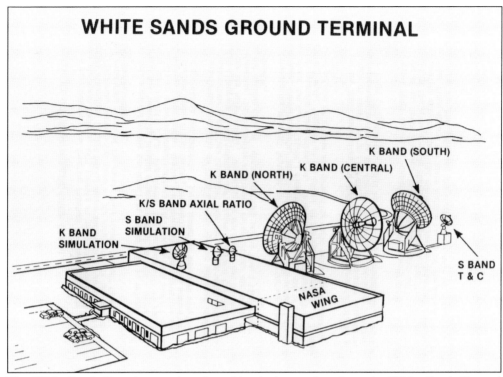

WHITE SANDS GROUND TERMINAL

K BAND (SOUTH)

K BAND (CENTRAL)

K BAND (NORTH)

K/S BAND AXIAL RATIO

S BAND SIMULATION

K BAND SIMULATION

S BAND T & C

NASA WING

The White Sands Complex consists of two identical ground stations, the White Sands Ground Terminal (WSGT) and Second TDRSS Ground Terminal (STGT). To create a link to its Southwest location, NASA renamed the two terminals from the local Tortugas tribal language. WSGT was designated "Cacique" (kah-see-keh), meaning "leader," and STGT was designated Danzante (dahn-zahn-teh), meaning "dancer." (Courtesy of NASA.)

While two differing acronyms are used, both are pointedly descriptive of the system. The TDRS acronym describes the Tracking and Data Relay Satellite, the on-orbit spacecraft. TDRSS, or Tracking and Data Relay Satellite System, includes the space segment, ground station, and communication links between all. (Courtesy of NASA.)

This image shows an artist's conception of the first-generation TDRS satellite on-orbit. First launched on STS-6 in 1983, the TDRS constellation was left with a sole satellite on orbit until 1988, after NASA's return to space after the *Challenger* accident. Even TDRS-1 had its trials getting positioned. After its inertial upper-stage booster blew up, it inched its way out to geosynchronous orbit (22,300 miles) using fist-sized maneuvering thrusters. (Courtesy of TRW, Inc.)

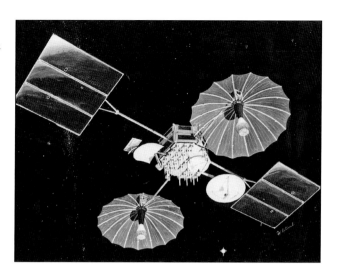

In its folded configuration, a second-generation TDRS awaits loading on its Atlas booster. The first-generation spacecraft (the standard design of the first six, plus a modified seventh) were unique; the second generation used a commonly used communications satellite bus to keep costs down. Awaiting launches in 2013 and 2014 respectively, TDRS-K and TDRS-L will form the third generation of the constellation. A third, TDRS-M, has been ordered for a 2015 launch. (Courtesy of NASA.)

A second-generation TDRS rides aboard an Atlas V to its new home 22,300 miles away in orbit. While conceptually nothing more than repeaters in the sky, communication satellites have allowed connection to virtually any place on the globe. In fact, TDRS-1 and its highly inclined orbit allowed the first pole-to-pole call, connecting from McMurdo Station in Antarctica to an Iridium satellite phone in the Arctic in April 1999. (Courtesy of NASA.)

Remembered by history as the last flight of the space shuttle *Challenger* and carrying the first teacher in space, the STS-51-L mission also contained the second satellite in the TDRSS constellation. The loss of the shuttle was a dual tragedy for the US space program, foremost because of the horrific loss of life and secondly for the incomplete space communications constellation, vital to many national and civil programs. (Courtesy of NASA.)

An image of the Aerospace Data Facility–Southwest (ADF-SW) logo clearly shows the Zia symbol in its design. In 2008, the secretive National Reconnaissance Office (NRO) made the unprecedented move to declassify the existence of its Mission Ground Stations in the United States. Three locations were revealed (ADF-Colorado in Denver, ADF-East near Washington, DC, and ADF-Southwest near Las Cruces), along with two overseas locations (United Kingdom and Australia). (Courtesy of the National Reconnaissance Office.)

The only overt indication of the National Reconnaissance Office's presence is an "Adopt-a-Highway" sign off of Highway 70. According to declassified NRO records, the ADF-SW is "a multi-mission ground station responsible for supporting worldwide defense operations and multi-agency collection, analysis, reporting, and dissemination of intelligence information." It is one of three ADFs in the United States; these sites are responsible for command and control of the nation's reconnaissance satellites.

A present-day view shows the weather-beaten WSTF sign. The WSTF team has gone through many ups and downs in the decades after the end of Apollo, supporting high-visibility programs such as the DC-X Delta Clipper and the Constellation human spaceflight program. The follow-on to Apollo's legacy, Constellation was cancelled in 2010, leaving the US space program with no domestic manned spaceflight capability for the first time since 1961.

At 05:29:45 Mountain War Time on July 16, 1945, the human race entered the Atomic Age. Trinity was the first detonation of a nuclear device in history, as a precursor to using the device in combat. On August 6, 1945, a B-29 Superfortress dropped a nuclear device code-named Little Boy on Hiroshima, Japan. Three days later, on August 9, 1945, another device, named Fat Man, was detonated over Nagasaki, Japan. (Courtesy of the National Archives.)

The Trinity fireball is pictured at 16 milliseconds after the blast was initiated. The grounds beneath the blast were subject to intense heat, fusing the silica together and creating a new mineral dubbed trinitite. Most of the glassy material was bulldozed and disposed of, but rare pieces can still be seen at the Trinity Site today. However, removing trinitite from the site is considered a crime; local rock shops contain samples for purchase. (Courtesy of the National Archives.)

This aerial view shows Ground Zero. The dangers of prolonged radiation exposure were unknown to scientists after the Trinity blast, so access to the site was limited from 1945 to 1953. (Courtesy of the US Air Force.)

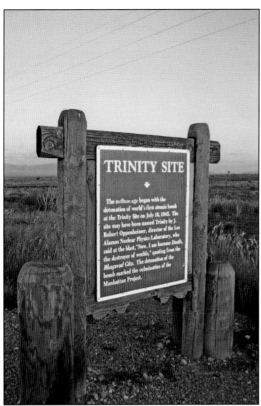

Trinity Site's New Mexico state historical marker along Highway 380 has seen its share of vandalism. With words scratched out or portions of text obscured, the sign takes the brunt of the protesting public's ire towards the unleashing of the nuclear genie.

The granite obelisk at Trinity's ground zero plainly states, "Trinity Site: Where The World's First Nuclear Device Was Exploded On July 16, 1945." (Courtesy of the US Air Force.)

SpaceShipTwo and its carrier craft, White Knight Two, are reflected in the window of the terminal hangar facility of the Spaceport America site near Upham, New Mexico. On October 17, 2011, the facility was renamed the Virgin Galactic Gateway to Space, after the spaceport's anchor tenant. (Courtesy of Cathy Harper.)

Members of the Project Bandaloop dance troupe perform along the glass surface of the Virgin Galactic Gateway to Space, located at Spaceport America. Virgin Galactic's enigmatic owner, Sir Richard Branson, also took to the aerial acrobatics show. (Courtesy of Cathy Harper.)

New Mexico Spaceport Authority executive director Christine Anderson (second from left), sits across from Sir Richard Branson (third from right) during the formal christening of the Virgin Galactic Gateway to Space. Anderson was the founding director of the Space Technology Directorate at Kirtland Air Force Base, New Mexico. A recognized space expert, she is responsible for the development and operation of the world's first commercial spaceport. (Courtesy of Cathy Harper.)

Second man on the moon astronaut Edwin "Buzz" Aldrin (left) presents a plaque to Sir Richard Branson during christening ceremonies at Spaceport America. Virgin Galactic's use of the Spaceport America facilities has promise to reinvigorate interest in space in southern New Mexico, as well as the global commercial space industry. (Courtesy of Cathy Harper.)

Virgin Mothership Two (VMS Two), christened Eve after Sir Richard Branson's mother, is seen during the dedication of the Virgin Galactic Gateway to Space on October 17, 2011. The craft is designed to carry its payload, SpaceShipTwo, to an altitude of 60,000 feet. After reaching its terminal altitude, SpaceShipTwo will uncouple from VMS Two and rocket off along its suborbital path. Branson's mother is artistically portrayed on the nose of the craft. (Courtesy of Cathy Harper.)

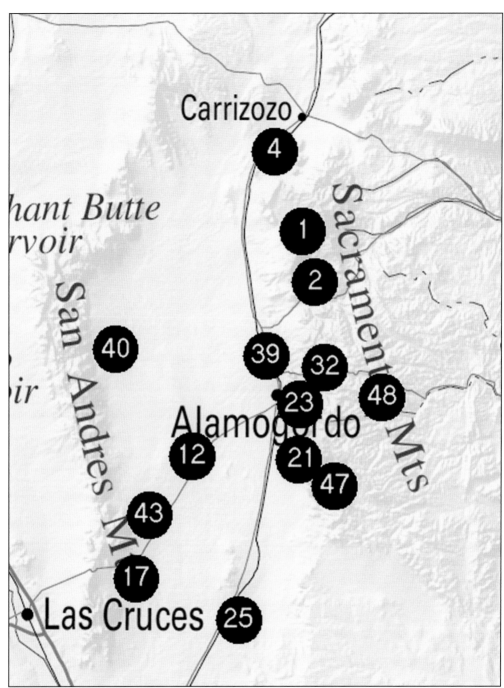

Tularosa Basin sites include: 1. Wizard's Roost Observatory; 2. Wally's Dome Observatory; 4. Three Rivers Petroglyphs Site; 12. Holloman Air Force Base; 17. White Sands Missile Range; 21. National Solar Observatory/Sacramento Peak; 23. New Mexico School for the Blind and Visually Impaired; 25. McGregor Range; 32. Tzec Maun Observatory; 39. New Mexico Museum of Space History; 40. White Sands Space Harbor; 43. High Energy Laser Systems Test Facility; 47. Apache Point Observatory; and 48. Comet Hale Bopp Codiscovery.

Six

TULAROSA BASIN

The Tularosa Basin has provided wonder and amazement since the first humans arrived there roughly between 9000 and 7500 BCE. The Mogollon people inhabited the area for a time and later abandoned it. The first interest by Europeans occurred in the mid-1800s, when US Army scouts ventured through. Harsh weather and scarcity of animals made living hard for follow-on farmers and ranchers. Towns such as Alamogordo were established as railheads for cattle, while others like Orogrande became gold rush sites, then ghost towns. The low population and wide spaces between two mountain ranges were increasingly attractive to the federal government's desires to create a missile testing range in the continental United States.

Holloman Air Force Base and White Sands Missile Range combined ranges in 1952, creating a national test range for the Department of Defense. Ostensibly run by the US Army, the range continually sees weapons tests, from surface-to-air missiles to drone aircraft destruction.

Astronomical observation sites in the Tularosa Basin encompass the entire spectrum of activities. The earliest archaeoastronomy observation sites in the state have been discovered in the Sacramento Mountains. Nearby prehistoric petroglyphs at the Three Rivers site graphically detail the views of ancient man, including thoughts on celestial bodies such as stars, moons, and galactic designs. This is not surprising, since the skies above southern New Mexico are perfect for stargazing, a fact noted by ancient sky gazers on through to present-day astronomers (and the occasional intelligence collection organization).

The many telescopes of Sunspot and Apache Point provide a wealth of knowledge to university researchers, while smaller endeavors such as Earthrise Institute and the Tzec Maun Foundation give similar views of the heavens to students and amateurs around the world.

The snow-white gypsum dunes of the White Sands National Monument (WSNM) reflect blinding sunlight into the camera. Studies of WSNM from earth-bound geologists allow testing of hypotheses of Martian geologic processes. Meridiani Planum on Mars contains similar features to WSNM, such as varied temperature cycling, presence of humidity within the sands, and frost growth depending on the season. These similarities have helped NASA mission planners choose future Martian observation sites. (Courtesy of Erin E. Gaberlavage.)

The Wizard's Roost site was accidentally discovered during a cultural survey for the Bureau of Indian Affairs in 1977. After assessment by local experts, including Jack Evans, retired director of Sunspot Solar Observatory, it was determined that Wizard's Roost was an azimuthal back sight for a solstice-centric observation post. One alignment confirmed tracing the heliacal rise of Sirius, the "Dog Star," dating the site to approximately 100 BCE to 900 AD. (Courtesy of Peter Eidenbach.)

The winter solstice sun rises above the Pecos Plains, east of the Sacramento Mountains, perfectly aligning itself within the rocks at the Wally's Dome archaeoastronomy site. Ancient mountain dwellers used solstice observation posts such as these to determine the precise planting and harvest seasons. Seasonal weather here is distinct, with harsh heat and cold temperatures. Having a slip at the start of planting season would have been disastrous to any culture surviving primarily on agrarian means. (Courtesy of Peter Eidenbach.)

This photograph of Wizard's Roost was taken during investigation of its possible summer sunset alignment. After seasonal confirmation of Wizard's Roost's raison d'être, deductive reasoning along with a conspiracy-fueled local mineral prospector led the cultural survey team to another nearby archaeological site, dubbed Wally's Dome. (Courtesy of Pete Eidenbach.)

Peter Eidenbach's artistic conception of the solstice sunrise over Wizard's Roost shows the observation room reconstructed from the rocky remnants. (Courtesy of Peter Eidenbach.)

A close-up shows the assembled room at Wizard's Roost. No ancient artifacts were found at the site. However, its construction leaves no doubt of human occupation. (Courtesy of Peter Eidenbach.)

Researchers warm near a campfire, waiting for sunrise on winter solstice in the Sacramento Mountains. Once the site was hypothesized to be an ancient solstice calendar, the theory needed proof. (Courtesy of Peter Eidenbach.)

The sun aligns itself through a rock structure at Wally's Dome during summer solstice. (Courtesy of Peter Eidenbach.)

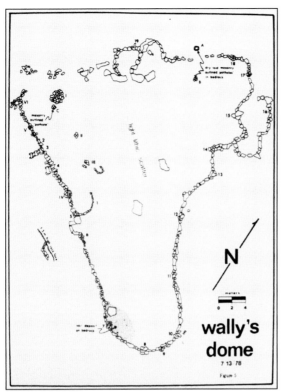

The site at Wally's Dome was planned out by its inhabitants with the inclusion of a rock barrier around the perimeter. (Courtesy of Peter Eidenbach.)

The summer sunrise over the southern New Mexico skies is viewed by an anthropological research team testing the hypothesis of Wally's Dome as a solstice observatory. (Courtesy of Peter Eidenbach.)

Mark Wimberly looks at non-randomly distributed glacial boulders. Vandals have destroyed many portions of Wizard's Roost. Sadly, Wimberly lost his life in a helicopter crash above Wizard's Roost in 1980. (Courtesy of Peter Eidenbach.)

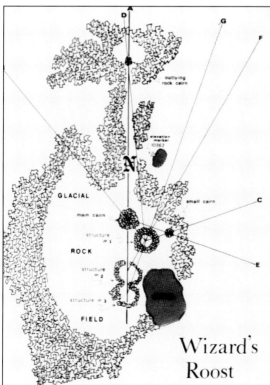

The site layout at Wizard's Roost included three circular rooms created by stacked rocks from a glacial debris field. The rooms provided environmental protection from wind, rain, and snow for the prehistoric observer corps. Additional sighting means were provided by human-made cairns, or intentional piles of stacked stones. The rooms and cairns provided bore-sight alignment for solstice observations. (Courtesy of Peter Eidenbach.)

A coiled serpent petroglyph at Three Rivers is just one of hundreds at the site. In the distance, Sierra Blanca displays a fresh coat of snow. Wizard's Roost and Wally's Dome are in the mountains beyond. (Courtesy of Erin E. Gaberlavage.)

Facing northwest of Three Rivers, the Oscura Mountains shield the location of Trinity Site. Note the chipped petroglyph in the lower right corner. An amazing number of space-related sites are within a few hours of this spot. (Courtesy of Erin E. Gaberlavage.)

The STS-3 mission lands at Northrup Strip, White Sands Missile Range, New Mexico, on March 30, 1982. The landing strip at Edwards Air Force Base, California, was flooded, forcing NASA to reroute *Columbia* to the alternate runway at WSMR. (Courtesy of NASA.)

Two NASA T-38 chase planes follow *Columbia* as it touches down at White Sands' Northrup Strip. Northrup Strip had been used by NASA for shuttle approach and landing testing before STS-3's landing. (Courtesy of NASA.)

Locals view the space shuttle *Columbia* as it is readied to be lifted aboard its 747 carrier aircraft. Pres. Ronald Reagan signed Senate Bill 2373, renaming the landing site White Sands Space Harbor, on May 11, 1982. (Courtesy of NMMSH.)

This photograph is a pilot's-eye view of White Sands Space Harbor (WSSH) during approach and landing. During the space shuttle program, WSSH was used extensively by the Shuttle Training Aircraft (STA). The STAs, modified versions of the Gulfstream II, were designed to mimic the shuttle's aerodynamic behavior inside an atmosphere. The shuttle was nicknamed "the Flying Brick," so pilots required additional practice landing. (Courtesy of the White Sands Test Facility Public Affairs Office.)

Col. John P. Stapp rides the Sonic Wind No. 1 test sled down the Holloman High Speed Test Track with the water brakes forcing huge plumes into the air. Colonel Stapp became the fastest man alive on his 29th (and final) sled test. His speed of 632 miles per hour was faster than a speeding bullet and subjected his body to 46.2 Gs of force. (Courtesy of the US Air Force.)

An HH-43 Pedro helicopter snatches an incoming Discoverer/Corona retrieval bucket in mid-air. The first few Discoverer/Corona recoveries were unsuccessful, threatening to derail the covert program, so three tests of the parachute-bucket combination took place over Holloman Air Force Base in the late 1950s. (Courtesy of the US Air Force.)

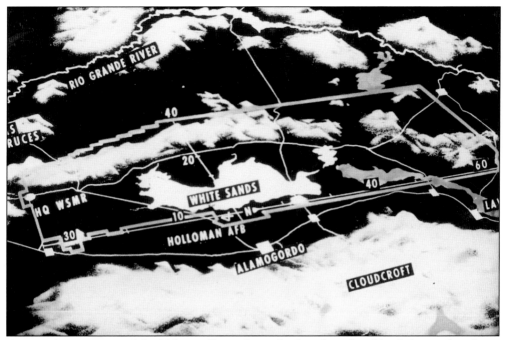

An artist's conception of the Tularosa Basin shows the length and width of the combined test range of WSMR and Holloman Air Force Base. (Courtesy of the US Air Force.)

A Project Moby Dick balloon is inflated at the balloon launch pad on Holloman Air Force Base. New Mexico is not a stranger to balloon research activities, starting with drop tests of anthropomorphic dummies in the 1950s, Project Manhigh and Excelsior manned balloon tests, and top-secret photographic reconnaissance projects like Moby Dick. (Courtesy of the US Air Force.)

An unidentified Air Force member interviews George House (right) while Lt. Col. Wayne O. Mattson (retired) looks on. The NMMSH received many test articles from Holloman's range, such as a TM-61 Matador cruise missile, that sit in the downtown lot awaiting refurbishment. Lieutenant Colonel Mattson volunteered at the museum for many years, having worked on many of the programs tested out on the White Sands/Holloman test ranges. (Courtesy of the US Air Force.)

A chimpanzee from Holloman Aeromedical Research Laboratory trains on space equipment. The first chimp in space, Ham, was named after Holloman Aeromedical and was actually known as Number 65 until after his space flight. (Courtesy of the US Air Force.)

The 4th Satellite Communications Squadron (Mobile) logo is shown here. The unit moved to Holloman Air Force Base in 1986 and had one of the most important missions in the US Air Force during the Cold War. When given direction from the Joint Chiefs of Staff, the unit would disperse its vehicles throughout three Southwest states to provide early warning data to the Pentagon in the event of a surprise Soviet nuclear missile strike.

The 4th Satellite Communications Squadron (Mobile) vehicles are photographed in deployed configuration. The unit's mission was declassified in 1992, after the end of the Cold War. While the mission of missile warning was still imperative for national security, the covert "midnight movement" tactics were less needed. In 1996, the unit gave up its trucks and transferred them to the 137th Space Warning Squadron, part of the Colorado Air National Guard. (Courtesy of TRW, Inc.)

The logo of the 781st Test Squadron, better known as RATSCAT (Radar Target Scatter), shows the Latin motto for "Unseen, Unhurt," emphasizing the effect of low-observable technology for the war fighter. (Courtesy of Air Force Historical Research Agency.)

A missile test body sits on the RATSCAT range. The test site places an aircraft, spacecraft, or missile on a pillar and bombards it with radar energy, attempting to determine its radar cross-section. (Courtesy of the US Air Force.)

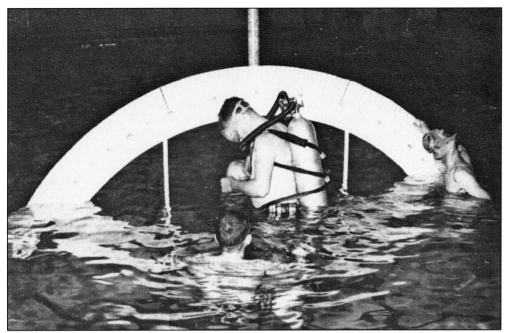

Dr. Grover D. Schock, the first person to receive a doctorate in space biology from the University of Illinois, pondered how astronauts would react in zero gravity in regards to orientation. Using the pool at the nearby New Mexico School for the Blind and Visually Impaired, members of the Holloman Aeromedical Research Laboratory fabricated a seat on a long pole and then spun a test subject in the water to test for disorientation. (Courtesy of the US Air Force.)

Happy pool participants splash inside the New Mexico School for the Blind and Visually Impaired's aquatic facilities. While the school's contributions may seem minor in the scope of space history, the proximity of Holloman Aeromedical Research Laboratory, testing at White Sands, and the zero gravity studies there created a mecca of space research. (Courtesy of the New Mexico School for the Blind and Visually Impaired.)

The High Energy Laser Systems Test Facility (HELSTF) has housed directed energy weapons research since the 1980s with the Strategic Defense Initiative (Star Wars) program.

A test model of the Apollo Launch Escape System sits atop an engineering boilerplate of the Command Module. WSTF used the nearby WSMR range facilities to test the LES atop a Little Joe II rocket on May 13, 1964. (Courtesy of WSMR Museum.)

A Little Joe II lifts off from WSMR during a test firing of the Apollo LES on May 13, 1964. The test was a rousing success and provided good data for system performance. (Courtesy of WSMR Museum.)

A V-2 arrives at Launch Complex 33 on May 10, 1946. The missile sits upon its transport, called a *Meillerwagen*. Only three such vehicle transports are known to still exist out of 200 manufactured. (Courtesy of WSMR Museum.)

V-2 No. 59 was launched on May 20, 1952. Note the paint scheme with "Buy Bonds" stenciled on the missile's midsection. (Courtesy of WSMR Museum.)

Capt. Andrew F. Breland (center), chief ranger Hugh BoZeus (right), and an unidentified soldier pose next to the White Sands National Monument entry sign. Such public relations photographs are vital to continuing the military and National Park Service's positive costewardship of the area. (Courtesy of WSMR Museum.)

Retrieval of a spent rocket casing on White Sands Missile Range is undertaken by land and air by both National Park Service personnel and the US military. (Courtesy of WSMR Museum.)

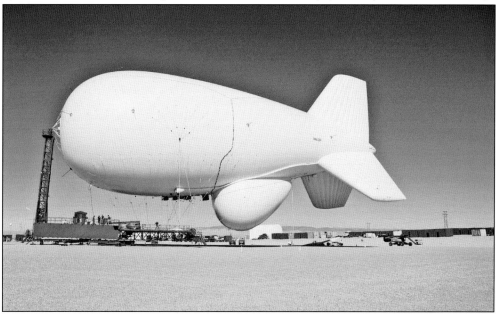

Capitalizing on a concept known as "near space," an aerostat at WSMR provides overhead communications linkage of Army ground forces back to their headquarters. Used extensively during the Army's Network Integration Evaluations (NIEs), the aerostat gives near constant overwatch during the tests, providing continuous communications. Geosynchronous communications satellites cost millions of dollars to build, launch, and operate. In contrast, aerostats are cheaper, more localized, and have higher digital data rates. (Courtesy of the Department of the Army.)

A High Mobility Artillery Rocket System (HI-MARS) launch takes place on WSMR. The physics of military-oriented rocket artillery and researchers' sounding rockets are exactly the same: the rocket gets launched, follows a ballistic trajectory, and comes back down to earth. WSMR's ability to use experience and knowledge gained from both military and research launches allows cross-flow between the two for the benefit of both. (Photo courtesy of the Department of the Army.)

An unidentified US Army soldier watches over a mock village, simulating civil operations being used in Afghanistan. After September 11, 2001, the military maneuver spaces at McGregor Range were adapted to assisting the actions in the Global War on Terror. Pseudo–Middle Eastern villages sprang up at different locations in the Tularosa Basin to assist deploying Army personnel in training how to handle specific situations. (Courtesy of the Department of the Army.)

The darling of Desert Storm, the MIM-104 PATRIOT, is launched at McGregor Range. Created as SAM-D (Surface-to-Air Missile–Developmental), the missile system was renamed Phased Array Tracking Radar to Intercept on Target, or PATRIOT, in 1976. Originally designed as an anti-aircraft system, the current variant has the ability to engage tactical ballistic missiles. Still in use by the United States and 14 allies, PATRIOTs are routinely flight-tested at McGregor Range. (Courtesy of the Department of the Army.)

A Stinger missile is launched from a Humvee-mounted Avenger missile system during maneuvers at McGregor Range. The background shows the close distance to the nearby Sacramento Mountains. National Guard, Army Reserve, and active-duty Army units come to McGregor to live-fire, an activity they are unable to do in their home states. Travelers along Highway 54 will see a menagerie of Army systems on maneuvers on any given day. (Photograph by Sfc. Alejandro Sias, courtesy of Department of Defense.)

The Sunspot Astronomy and Visitor Center is located at the southern end of New Mexico Scenic Byway 6563. A joint venture between the National Solar Observatory (NSO), the US Forest Service, and Apache Point Observatory, it is open every day from 9:00 a.m. to 5:00 p.m. from March to January for self-guided tours. Exhibits within the center highlight NSO activities, the work done at Apache Point, and the stewardship of the US Forest Service.

The Richard B. Dunn Solar Telescope was the world's top-of-the-line optical solar telescope when inaugurated in 1969; still imposing today, the telescope continues to be a premier site for solar observations. The aboveground structure towers 136 feet but is dwarfed by the 228 feet of underground structure. The entire optical system is a rotating structure, with over 200 tons of equipment suspended from the top by a mercury float bearing.

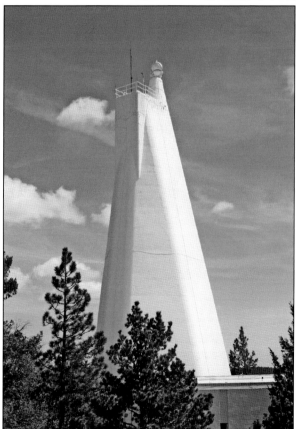

The Hilltop Dome was built to take patrol images of the sun. It contains equipment that archives observations of the sun, from visible light to the edge of the spectrum, showing the interaction of hydrogen atoms (6563 Angstroms). The archived record of sun observations by the Hilltop Dome runs over decades. The dome is not open to the public, but up-to-date images of the sun are broadcast inside the Dunn lobby.

The Grain Bin Dome telescope was built in 1950 as the first telescope at Sacramento Peak. The structure was purchased out of a Sears catalog and modified by on-site Air Force machinists to rotate on its axis. The six-inch (0.15-meter) telescope monitored solar prominences from 1951 through 1963. It continued work observing coronal observations until 1969. Today, it is home for a small telescope for local residents' viewing pleasure.

The Evans Solar Facility, housed in the Big Dome, is actually two telescopes in one: a 16-inch coronagraphic telescope mapping magnetic changes in the sun's outer atmosphere and a special set of 12-inch coelostat optics that "squint" so the sun appears as a pinpoint. The facility is mostly used to look at the corona, the faint outermost layer of the sun, but is also used to investigate solar flares.

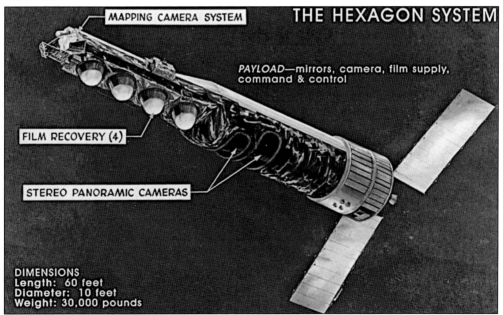

An artist's conception shows the Keyhole-9 (KH-9) Hexagon spy satellite. Hidden from the public eye for four decades, the KH-9 was declassified in 2011 by the National Reconnaissance Office. One interesting note revealed the KH-9 cameras were taken to the US Air Force's Celestial Calibration Site in Cloudcroft, New Mexico, for testing. Being 9,000 feet above sea level and away from urban lights, the location was perfect for surreptitious stargazing. (Courtesy of the National Reconnaissance Office.)

The coma from Comet Hale-Bopp is clearly seen during this morning-sky photograph taken on March 21, 1997. The codiscoverer of the comet, Dr. Alan Hale, furthered the impact astronomy has on the public by starting Earthrise Institute in 1993. The mission of Earthrise Institute is to use astronomy, space, and other related endeavors as a tool for breaking down international and intercultural barriers and for bringing humanity together. (Courtesy of Earthrise Institute.)

Comet Hale-Bopp is seen above a launch complex at WSMR prior to launch of an observation mission to study the comet. Earthrise Institute maintained a dark-sky site near Cloudcroft until 2010. It continues to support international astronomy through goodwill trips to view celestial events, such as solar eclipses in Iran in 1999, Zimbabwe in 2001, and most recently Australia in November 2012. (Courtesy of Earthrise Institute.)

Apache Point, near Sunspot, is a bit less visitor-friendly than its neighbor. The grounds include a short walk between telescopes, but few inside public viewing opportunities while daily research is conducted.

Students from the NMMSH's Space Academy view Apache Point's 3.5-meter telescope. (Courtesy of NMMSH.)

The Tzec Maun Foundation has multiple telescope farms in New Mexico and Australia. The telescopes are attached to charge-coupled device (CCD) cameras that are digitally connected to the Internet, allowing students and researchers around the world access. Using open-source software on desktop computers, users can enjoy quality images from professional-grade instruments viewing from some of the best dark-sky sites in the world. (Courtesy of the Tzec Maun Foundation.)

The Tzec Maun Cloudcroft site was originally a US Air Force astronomical installation. The construction of the facility was completed in 1962. The dome was 50 feet in diameter, allowing the telescope to be positioned without interference, and the base of the dome was four stories high to place the telescope above local ground effects. The Air Force's first operational electro-optical space surveillance sensor obtained first light in the late 1960s. (Courtesy of the Tzec Maun Foundation.)

Messier object no. 51 was captured during initial testing of Tzec Maun's one-meter telescope at Cloudcroft, New Mexico. The foundation's name comes from the ancient Mayan name for the planet Mercury, Tzec Maun. The name translates as "Skull Owl," a symbol of intellectual balance. (Courtesy of the Tzec Maun Foundation.)

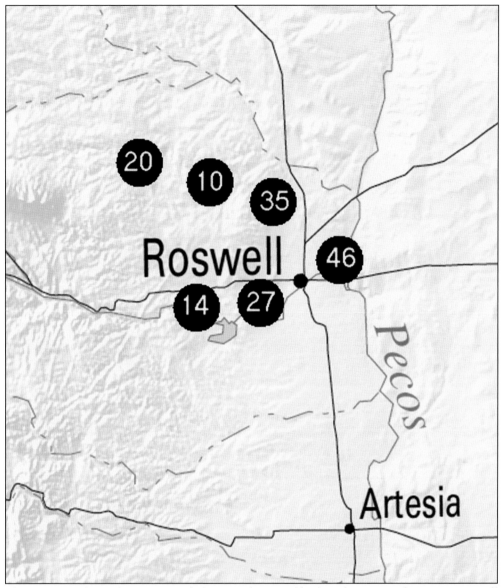

The Space Trail's southeast sites include: 10. Robert Goddard Rocket Research Site; 14. Roswell Army Air Field/Walker Air Force Base; 20. Alleged UFO Crash Sites; 27. Atlas Missile Silos; 35. Roswell Museum and Art Center; and 46. International UFO Museum and Research Center.

Seven

SOUTHEAST

The plains of southeastern New Mexico are bordered by the Sacramento Mountains on the western side, bisected by the Pecos River, and continue flat and straight through to the Texas border. Space-related activities here capitalize on geographic features as well as the added benefits of low population density and "salt of the earth" people with a wonderful work ethic. In the early 1930s, Dr. Robert Goddard chose Roswell as his home while he developed the world's first liquid-fueled rockets. Mishaps from his rocket tests in Massachusetts (include an infamous launch and subsequent crash that started a blaze and had neighbors complain to the fire marshal) forced Goddard to find a better testing ground where his rockets could fly.

Nearly a decade after Goddard's rocket tests, one of the most contested events in modern history occurred on a ranch outside of the city of Roswell. Whether the Army Air Forces captured an unidentified flying object, or "flying saucer" in the vernacular of the times, is still in dispute over six decades later. Soon after the incident, Roswell Army Air Field moved into the post–World War II Atomic Age with blazing speed. Walker Air Force Base, as the airfield was renamed, became one of the Strategic Air Command's premier installations. Housing two thirds of America's nuclear triad, with nuclear-armed bombers and nuclear-tipped intercontinental ballistic missiles, Walker kept alert crews ready to go to war throughout the 1950s and 1960s, until its closure in 1967.

Regardless of the truth, Roswell has seen a large tourism trade develop around the UFO crash. The UFO hype first draws visitors in and then allows Roswell to showcase other amazing attractions such as the Roswell Museum and Art Center (displaying Dr. Goddard's work), the International UFO Museum and Research Center, and Roswell International Air Center.

The famous July 8, 1947, front cover of the *Roswell Daily Record* stated that Roswell Army Air Field had "captured" a flying saucer. Details within the article were taken from information released by 509th Bombardment Group intelligence officer Maj. Jesse Marcel. Other named individuals in the article would not figure so prominently in the building of the case of the Roswell Incident, instead echoing the names of Marcel and rancher Mac Brazel. (Courtesy of the *Roswell Daily Record*.)

The front cover of the *Roswell Daily Record* for July 9, 1947, displays an indirect retraction from Eighth Air Force commander Gen. Roger M. Ramey. Later reports would find Maj. Jesse Marcel posing with debris from a weather balloon for local newspapers. Ordered by his superiors, the photographs destroyed Marcel's credibility about the incident, forcing people to question his ability to distinguish between broken wood and aluminum foil and extraterrestrial ships. (Courtesy of the *Roswell Daily Record*.)

The International UFO Museum is a major attraction in downtown Roswell. Since it opened in 1992, the museum has seen thousands of curious people ranging from UFO fanatics to serious researchers. The archives of the museum hold artifacts and documents relating to thousands of UFO encounters around the world and the facility is considered a leader in the area of UFO research.

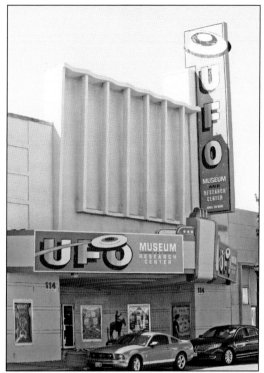

The author stands inside a diorama of the 1947 Roswell UFO crash. The background painting is similar to the results seen by rancher Mac Brazel when the debris was found. Brazel had noticed a cut in the earth and random bits of metal and other materials from a distance away, but the crash was not distinctive enough to warrant closer inspection until almost three weeks later.

A declassified photograph shows an aeroshell tested at White Sands Missile Range in 1967. The design was for the Viking program, to allow a lander to safely arrive on the red planet. Conspiracy theorists attempt to link the Roswell crash with technology for this and similar "flying saucer" designs. Even the Air Force's 1995 report on the Roswell crash would explain the incident as mistaken identity with top-secret programs. (Courtesy of the US Air Force.)

The 509th Bombardment Group sign outside of Roswell Army Airfield prominently displays the worldwide reach of the unit's B-29 atomic bombers. The connection between the Roswell crash and the 509th would jokingly fuel theories decades later that the present-day 509th Bomb Wing's B-2 stealth bombers were designed with or powered by alien spacecraft technology. (Courtesy of the Walker Aviation Museum Foundation.)

On September 18, 1942, AT-11 Kansan navigation trainers from the Roswell Army Flying School fly over New Mexico Military Institute (NMMI). The flight was dual-purpose: training for the Army Air Forces' aviation cadets and a public air display saluting the cadets of NMMI, one of six military junior colleges in the United States. (Courtesy of the Historical Society of Southern New Mexico.)

A B-47 Stratojet sits on the Walker Air Force Base ramp alongside a visiting B-29 Superfortress during the base's 1952 air show. The unit of the B-47 is unknown, as the 509th Bombardment Wing did not receive its Stratojets until 1955. (Courtesy of the Walker Aviation Museum Foundation.)

Visiting from the 127th Pilot Training Wing at Luke Air Force Base in Arizona, F-84B serial number 46-621 and its sister craft line the Walker Air Force Base ramp. Common practice for military aviation training includes flights to nearby Air Force installations. The practice continues at Roswell International Air Center with present-day military training flights. (Courtesy of the Walker Aviation Museum Foundation.)

B-52E serial number 56-0635 is refueled in-flight by a KC-135A Stratotanker, serial number 57-1467. Aircrews from the 6th Bombardment Wing at Walker Air Force Base flew both the B-52E and KC-135A models until the base's closure in 1967. American Cold War strategy positioned bomber and air refueling squadrons side-by-side at Strategic Air Command bases to guarantee maximum preparedness for nuclear combat. (Courtesy of the Walker Aviation Museum Foundation.)

An aerial view shows Walker Air Force Base before its closure in 1967. One of the most distinctive features of Strategic Air Command bases during the Cold War was the "Christmas tree" alert pads, seen at the lower center of the photograph. The design allowed a staggered flow of aircraft from the alert ramp to the runway. Present-day military aviation trainees enjoy using the 13,000-foot runway for takeoff and landing practice. (Courtesy of the Walker Aviation Museum Foundation.)

Dr. Robert Goddard (center) stands with Harry Guggenheim (left) and Charles Lindbergh (right) at the Roswell launch tower around 1935. With personal attention from "Lucky Lindy," Dr. Goddard received financial support from the Daniel and Florence Guggenheim Foundation for his rocketry testing. The subsequent paper, titled *Liquid Propellant Rocket Development*, became the foundational document for modern rocket designs, establishing Dr. Goddard as the "Father of Modern Rocketry." (Courtesy of the Roswell Museum and Art Center.)

Dr. Robert Goddard looks towards the launch gantry, telescope in one hand and ignition switch in the other. After the debacle of the paper *A Method of Reaching Extreme Altitudes* and a continuing apathy towards his rocketry work from the US government, Goddard became even more suspicious and hesitant to work with others. During the buildup to World War II, letters from German rocket scientists would arrive and go unanswered. (Courtesy of the Roswell Museum and Art Center.)

A statue of Goddard sits outside of the Roswell Museum and Art Center. On July 17, 1969, one day after the Apollo 11 moon launch, the *New York Times* stated, "Further investigation and experimentation have confirmed the findings of Isaac Newton in the 17th Century and it is now definitely established that a rocket can function in a vacuum as well as in an atmosphere. The *Times* regrets the error."

Robert Goddard (front right) and four colleagues carry a rocket body and nose cone to their launch site. (Courtesy of the Roswell Museum and Art Center.)

A proud inventor stands beside his invention. Robert Goddard's impact on rocketry was best summed up by Dr. Wernher von Braun, father of the V-2 rocket, after the end of World War II. When asked about Goddard's work by American reporters, von Braun quipped, "Don't you know about your own rocket pioneer? Dr. Goddard was ahead of us all." (Courtesy of the Roswell Museum and Art Center.)

A replica of Goddard's gantry is displayed in Roswell, New Mexico. The gantry is surrounded by a quote recognizing October 19 as the day his dedication to rocketry became fixed. After climbing a cherry tree to cut some dead limbs, Goddard looked up at the sky and was transfixed, "[imagining] how wonderful it would be to make some device which had even the possibility of ascending to Mars."

An Atlas-F intercontinental ballistic missile is readied for a test launch at Vandenberg Air Force Base, California. Since it was not prudent to test launch missiles over American soil, launch crews from Walker Air Force Base and other Strategic Air Command installations traveled to California for test launches over the Pacific Ocean. Nuclear deterrence required near-perfect vigilance, and Atlas crewmembers were no exception. After their California "vacations," crews soldiered back to New Mexico for nuclear alert. (Courtesy of 90th Missile Wing History Office.)

Outside of Roswell, Atlas missile site 579-10 has its enclosure doors opened, exposing the launch facility interior to the elements. This open-door sight was common on later missile systems, as their deactivation was monitored by Soviet surveillance satellites. Atlas was voluntarily withdrawn from service as safer, more reliable missile systems such as the Minuteman came online. The shorter ranges of newer systems eliminated Walker Air Force Base as a possible launch location.

A missile transporter truck takes an Atlas missile to emplacement at one of the 12 launch sites surrounding Roswell. The Atlas, with its liquid-fueled design, represents the pinnacle of Dr. Robert Goddard's rocket designs from nearly three decades earlier. Sadly, Goddard would not live to see the distributed uses of his work around his adopted home, ranging from nuclear weapon delivery vehicle to space launch vehicle for manned missions. (Courtesy of the 579th Strategic Missile Squadron Association.)

Humankind's space history melds beautifully in the skies and lands of New Mexico, with a waxing gibbous moon hanging in the evening sky, keeping watch over the sand dunes of White Sands National Monument. These inspiring sights have been commonplace since the dawn of man; through a fortuitous mixture of people, resources, and spirit, New Mexico became the perfect place to help man observe and later reach for the stars. (Courtesy of Erin E. Gaberlavage.)

About the New Mexico Museum of Space History

The mission of the museum is to educate the people of New Mexico and our visitors from around the world in the history, science, and technology of space. The museum stresses the significant role that the state of New Mexico has played in the development of the US space program through collecting, preserving, and interpreting significant artifacts relevant to the history of space.

The New Mexico Museum of Space History is one of 15 divisions of the New Mexico Department of Cultural Affairs. The museum is composed of:

The Museum of Space History, which contains exhibitions ranging from Robert Goddard's early rocket experiments near Roswell to a mock-up of the International Space Station.

The International Space Hall of Fame, which commemorates the achievements of men and women who have furthered humanity's exploration of space.

The John P. Stapp Air & Space Park, which displays larger exhibits such as the Apollo program's Little Joe II rocket and the rocket sled that "Fastest Man Alive" Stapp rode to 632 miles per hour.

Daisy Track, which commemorates aeromedical and space-related tests that were crucial in developing components for NASA's Mercury orbital flights and Apollo moon landings.

The Clyde W. Tombaugh IMAX Theater and Planetarium, the only such theater in southern New Mexico.

The Astronaut Memorial Garden, a tribute to the Apollo 1 and space shuttle *Challenger* and *Columbia* astronauts.

The Hubbard Space Science Research Building, home to the museum's new archives and library. Researchers and students will find an academic-based collection of New Mexico space history, Holloman Air Force Base and White Sands Missile Range information and photographs, and NASA publications, photographs, and collections.

The Museum Support Center, where museum employees and volunteers conserve and restore the many large artifacts exhibited at the museum.

The museum is open every day of the year except Christmas and Thanksgiving from 9:00 a.m. to 5:00 p.m. It is located at the top of Highway 2001 in Alamogordo, New Mexico.

DISCOVER THOUSANDS OF LOCAL HISTORY BOOKS
FEATURING MILLIONS OF VINTAGE IMAGES

Arcadia Publishing, the leading local history publisher in the United States, is committed to making history accessible and meaningful through publishing books that celebrate and preserve the heritage of America's people and places.

Find more books like this at
www.arcadiapublishing.com

Search for your hometown history, your old stomping grounds, and even your favorite sports team.

Consistent with our mission to preserve history on a local level, this book was printed in South Carolina on American-made paper and manufactured entirely in the United States. Products carrying the accredited Forest Stewardship Council (FSC) label are printed on 100 percent FSC-certified paper.

MADE IN THE
USA